Group Theory

Group Theory

Dr. Vijay Kumar Singh

SHREE PUBLISHERS & DISTRIBUTORS
NEW DELHI-110 002

Edition : 2008
Reprint : 2009

Published by :
SHREE PUBLISHERS & DISTRIBUTORS
22/4735 Prakash Deep Building,
Ansari Road, Darya Ganj,
New Delhi–110 002

ISBN : 978–81–8329–258–0

Printed by :
Vikas Computers & Printers
Delhi–110032

Preface

Group theory is that branch of mathematics concerned with the study of groups. There are three historical roots of group theory: theory of algebraic equations, number theory and geometry. Euler, Gauss, Lagrange, Abel and Galois were early researchers in the field of group theory.

Groups are used throughout mathematics and the sciences often to capture the internal symmetry of other structures in the form of automorphism groups. An internal symmetry of a structure is usually associated with an invariant property, and the set of transformations that preserve this invariant property, together with the operation of composition of transformations, form a group called a symmetry group.

Group theoretical principles are an integral part of modern physical sciences. In chemistry, groups are used to classify crystal structures, regular polyhedra and the symmetries of molecules. In physics, groups are important because they describe the symmetries which the law of physics seem to obey. This self-contained work, written for advanced undergraduate-level and graduate-level chemistry students, clearly and concisely introduces the subject of group theory and demonstrates its applications to chemical problems. Students and professionals in the fields of physical chemistry and chemical physics will find this work of utmost use.

While writing this book, I have made references from various sources and have freely used the writings of outstanding scholars and researchers. I, hereby, acknowledge their contribution with sincerity and gratitude.

I am thankful to the publisher for bringing out this book within a very short period.

Dr. Vijay K. Singh

Contents

Preface (*v*)

1. Introduction of Group Theory 1
2. Molecular Symmetry and Chemical Bonding 29
3. Vibrational Modes 101
4. Free Groups and Presentations 109
5. The Molecular Symmetry Group 164
6. Vibrational Spectroscopy 197
7. Group Theory and Quantum Mechanics 223
8. Recent Advances in Perturbation Theory 232
9. Classification of Electronic States of Acetylene 246
10. Structure and Bonding in Main Group Chemistry 273

1

Introduction to Group Theory

A complex number is just a pair, $z = (a, b)$ of real numbers. We usually write this pair in the form $z = a + ib$, where the "+" and "i" are just decorations (for now). The number a is called the real part of z, while b is called the imaginary part of z. We denote the set of all complex numbers by c. Note that we can represent any complex number $z = a + ib \in c$ by a point in the plane.

Examples 1. In class of complex numbers and their locations in the plane

Definition 2. We define addition and multiplication of complex numbers as follows:

(1) $(a + ib) + (c + id) = (a + c) + i(b + d)$

(2) $(a + ib)(c + id) = (ac - bd) + i(ad + bc)$

In other words, we add complex numbers by adding their real and imaginary parts, and multiply them by treating i as a square root of -1.

Notation 3. We use the following shorthand notation:

$$a + i \cdot 0 = a \text{ (i.e., } (a, 0) = a)$$

$$0 + ib = ib \text{ (i.e., } (0, b) = ib)$$

$$0 + i1 = i \text{ (i.e., } (0, 1) = i)$$

Then we see that the sum of a and ib is indeed the single complex number $a + ib$. In other words, we can now think of $a + ib$ as a sum rather than as a wild and crazy way of writing (a, b).

(1) $i^2 = -1$ (Check it and see; remember that i is just shorthand for the complex number $+i1$)

(2) For every complex number z, we have $1 \cdot z = z \cdot 1 = z$

(3) Addition and multiplication of complex numbers obey the same rules (commutativity, associativity, distributive laws, additive identity, multiplicative identity) as the real numbers. We will see of that there are also inverses.

Examples 4. Illustrating the geometry of addition and multiplication: In class.

Definition 5. The magnitude of the complex number $z = a+ib$ is given by the formula

$$|z| = \sqrt{a^2 + b^2}$$

(This is just its distance from the origin.) Also, we define:

$$\bar{z} = a - ib,$$

called the complex conjugate of z.

Examples in class

Now we notice that:

$$z\bar{z} = |z|^2$$

In other words:

$$z \cdot \frac{\bar{z}}{|z|^2} = 1$$

But this says that $\bar{z}\,|z|^2$ is the multiplicative inverse of z. In other words,

$$z^{-1}\ \frac{\bar{z}}{|z|^2}$$

Examples in class

We now look at the polar form, and we can write

$$z = r \cos \theta + ir \sin \theta = r(\cos \theta + i \sin \theta)$$

We also write this as $re^{i\theta}$.

Notes:

(1) If $z = re^{i\theta}$, then $r = |z|$.

(2) The identity $[r(\cos \theta + i \sin \theta)][s(\cos \phi + i \sin \phi)] = rs(\cos(\theta + \phi) + i \sin (\theta + \phi)$ translates to $re^{i\theta} + re^{i\theta} = re(^{i\theta + \phi})$, which is what we expect from the laws of exponents.

(3) Addition and multiplication of complex numbers. Also, this gives us the key to the geometric meaning of multiplication.

(4) If we multiply a complex number by itself repeatedly, we now get:

$$[r(\cos \theta + i \sin \theta)]^n = r^n(\cos n\theta + i \sin n\theta),$$

which is known as De Moivrés formula. We can use it to find nth roots of any complex number: take the nth root of the magnitude, and divide the angle θ by n. We can also divide θ + 2π by n to get another nth root, and θ + 4π ,θ +6π, etc. They start repeating when we get to θ + 2nπ. In other words:

There are n - 1 different nth roots of any complex number.

Examples 6.

A. Find all the 4th roots of i.

B. Find all the 5th roots of $32e^{i\pi}/3$.

C. Find all the roots of the equation $z^3 = i$.

If we choose $r = 1$ in De Moivrés formula, this places us on the unit circle, and we find all kinds of nth roots of 1.

Definition 7. The primitive nth root of unity is the complex number $\omega = e^{2\pi i/n}$. Note that all the other nth roots of unity are powers of ω. In other words, the nth roots of unity are:

$$1 = \omega^0, \omega, \omega^2, \ldots, \omega^{n-1}.$$

Sets, Equivalence Relations and Functions

A set is an undefined "primitive" notion. Intuitively, it refers to a collection of things called elements. If a is an element of the set S, we write $a \in S$. If a is not an element of the set S, we write $a \notin S$. Some important sets are:

Z, the set of all integers

N, the set of all natural numbers (including 0)

$Z+$, the set of all positive inegers

R, the set of all rational numbers

R, the set of all real numbers

C, the set of all complex numbers

We can describe a set in several ways:

(1) by listing its elements; e.g., $S = \{6, 66, 666\}$

(2) in the form $\{x \mid P(x)\}$, where $P(x)$ is a predicate in x, for instance

$$S = \{x \mid x \text{ is a real number other than } 6\}$$

or

$$T = \{x \in Z \mid x \text{ odd}\}$$

Note. Two sets are equal if they have the same elements. That is,

$$A = B \text{ means } x \in A \Leftrightarrow x \in B$$

Definitions 1. Let A and B be sets.

We say that A is a subset of B and write $A \subset B$ if $x \in A \Rightarrow x \in B$.

$A \cap B$ is the intersection of A and B, and is given by

$$A \cap B = \{x \mid x \in A \text{ and } x \in B\}$$

$A \cup B$ is the union of A and B, and is given by

$$A \cup B = \{x \mid x \in A \text{ or } x \in B\}$$

θ is the empty set; $\theta = \{x \mid F(x)\}$, where $F(x)$ is any false predicate in x, such as "$x = 3$ and $x \neq 3$"

$A - B$ is the complement of B in A, and is given by

$$A - B = \{x \mid x \in A \text{ and } x \notin B\}$$

$A \times B$ is the set of all ordered pairs,

$$A \times B = \{(a, b) \mid a \in A, b \in B\}$$

If $\{A_\alpha | \alpha \in \Omega\}$ is any collection of sets indexed by Ω, then:

is the intersection of the A_α and is given by

$$= \{x \mid x \in A_\alpha \text{ for all } \alpha \in \Omega$$

is the union of the A_α and is given by

Note. To prove that two sets A and B are equal, we need only prove that $x \in A \Leftrightarrow x \in B$. In other words, we must prove two things:

(a) $A \subset B$ (i.e., $x \in A$) $x \in B$)

(b) $B \subset A$ (i.e., $x \in B$) $x \in A$)

Lemma. The following hold for any three sets A, B and C and any indexed collection of sets $B_\alpha (\alpha \in \Omega)$

(1) Associativity

$$A \cup [(B \cup C) = (A \cup B) \cup C \; A \cap B \cap C)$$
$$= (A \cap B) \cap C$$

(2) Commutativity

$$A \cup B = B \cup A \; A \cap B = B \cap A$$

(3) Identity and Idempotent

$$A \cup A = A, \; A \cap A = A$$
$$A \cup \theta = A, \; A \cap \theta = \theta$$

(4) De Morgan's Laws

$$A-(B \cup C) = (A-B) \cap (A-C), \; A-(B \cap C) = (A-B) \cup (A-C)$$

Fancy Form:

$$A-\bigcap_{\alpha \in \Omega} B_\alpha = \bigcup_{\alpha \in \Omega} (A \cup B_\alpha), \; A \cap \bigcap_{\alpha \in \Omega} B_\alpha = \bigcup_{\alpha \in \Omega} (A \cap B_\alpha)$$

(5) Distributive Laws:

$$A \cup \bigcap_{\alpha \in \Omega} (B_\alpha) = \bigcap_{\alpha \in \Omega} (A \cup B_\alpha) \; (A \cap C), \; A \cap (B \cup C)$$

Fancy Form:

Proof. We prove (1), (2), (3) and a bit of (4) in class. The rest you will prove in the exercise set.

Definition 3. A partitioning of a set S is a representation of S as a union of disjoint subsets S_α, called partitions:

Examples 4.

A. The set Z can be partitioned into the odd and even integers.

B. $Z = 3Z \cup [\ (1 + 3Z)\ [\cup(2 + 3Z)$, where $m + 3Z = \{m + 3n \mid n \in Z\}$.

C. The set of $n \times n$ matrices can be partitioned into subsets, each of which contains matrices with the same determinant.

Definition 5. A relation on a set S is a subset R of $S \times S$. If $(a, b) \in R$, we write aRb, and say that a stands in the relation R to b.

Examples 6.

A. Equality on any set A

B. $\neq$ on any set A

C. $<$ on Z

D. $m \approx n$ if $m - n \in 3Z$, on Z

E. Row equivalence on the set of $m \times n$ matrices

F. Any partitioning of a set S gives one: Define aRb if a and b are in the same partition.

Definition 7. An equivalence relation on a set S is a relation $\approx$ on S such that, for all a, b and $c \in S$:

(1) $a \approx a$ (Reflexivity)

(2) $a \approx b \Rightarrow b \approx a$ (Symmetry)

(3) $(a \approx b$ and $b \approx c) \Rightarrow a \approx c$ (Transitivity)

Examples 8.

A. Equality on any set

B. Equivalence mod k on Z (in class)

C. Row equivalence on the set of $m \times n$ matrices

D. Any partition on S yields an equivalence relation

Definition 9. If $\approx$ is an equivalence relation on S, then the equivalence class of the element $s \in S$ is the subset

$$[s] = \{t \in S \mid t \approx s\}$$

Lemma 10. Let $\approx$ be any equivalence relation on S. Then

(a) If $s, t \in S$, then $[s] = [t]$ iff $s \approx t$.

(b) Any two equivalence classes are either disjoint or equal.

(c) The equivalence classes form a partitioning of the set S.

Theorem 11. (Equivalence Relations are Partitions)

There is a one-to-one correspondence between equivalence relations on a set S and partitions of S. Under this correspondence, an equivalence class corresponds to a set in the partition.

Examples 12.

A. Z/nZ is the set of equivalence classes of integers mod n. $[r] = [s]$ iff $r - s \in nZ$. We look at these equivalence classes explicitly in class.

B. Construction of the rationals define a relation on $Z \times Z^*$ by $(m, n) \sim (k, l)$ iff $ml = nk$. The quotient m/n is defined to be the equivalence class of (m, n).

Definition 13. Let A and B be sets. A map or function $f : A \to B$ is a triple (A, B, f) where f is a subset of $A \times B$ such that for every $a \in A$, there exists a unique $b \in B$ (that is, one and only one $b \in B$) with $(a, b) \in f$. We refer to this element b as $f(a)$. A is called the domain or source of f and B is called the codomain or target of f.

(1) We think of f a rule which assigns to every element of A a unique element $f(a)$ of B, and we can picture a function $f: A \to B$ as shown in the figure.

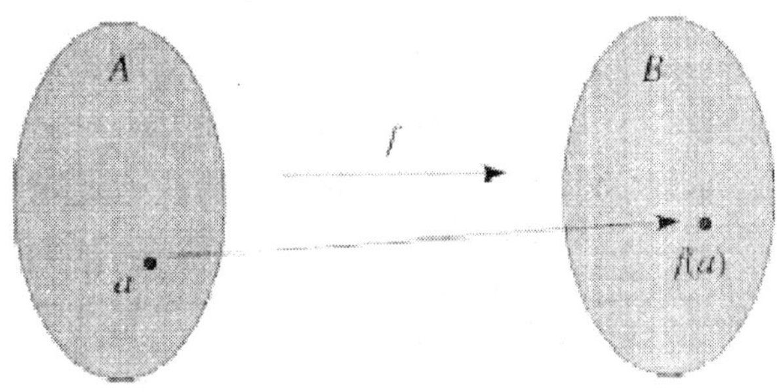

(2) The codomain of f is not the "range" of f; that is, not every element of B need be of the form $f(a)$.

(3) The sets A and B are part of the information of f. For instance, specifying f by saying only "$f(x) = 2x–1$" is not sufficient because we have not specified the domain and codomain. We should instead say something like this:

"Define $f: R \to R$ by $f(x) = 2x - 1$."

Examples 14. Some in class, plus:

A. If A is any set, we have the identity map $1_A : A \to A$; $1_A(a) = a$ for every $a \in A$.

B. If $B \subset A$, then we have the inclusion map : $B \to A$; $\iota\,(b) = b$ for all $b \in B$.

C. The empty map $\theta : \theta \to A$ for any set A.

Definition 15. Let $f : A \to B$ be a map. Then f is injective (or one-to-one) if

$$f(x) = f(y) \Rightarrow x = y$$

In other words, if $x \neq y$, then $f(x)$ cannot equal $f(y)$.

Examples 16.

A. $f : R \to R; f(x) = 2x - 1$ is injective.

B. $f : R \to R; f(x) = x^2 + 1$ is not.

C. The identity $1_A : A \to A$ is injective for every set A.

D. The inclusion $\iota : B \to A$ is injective for every set A and every subset $B \subset A$.

Definitions 17: Let $f : A \to B$ be a map, and let $C \subset A$. Then the image of C under f is the subset

$$f(C) = \{f(c) | \; c \in C\}$$

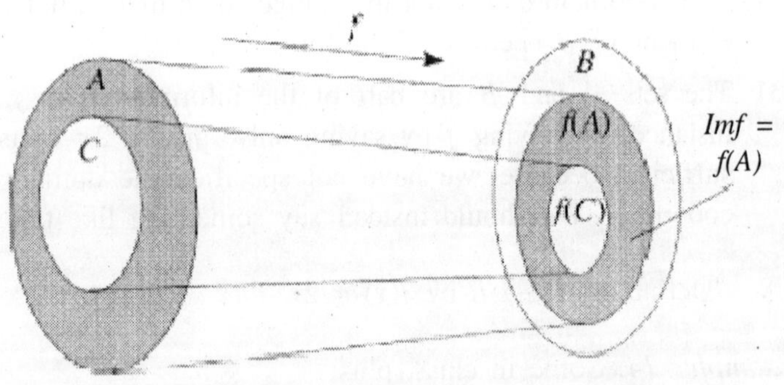

The image of f is defined as

$$\text{Im } f = f(A).$$

f is surjective (or onto) if Im $f = B$. In other words,

$$b \in B \Rightarrow \exists a \in A \text{ such that } f(a) = b$$

Thus, f "hits" every element in B (see figure).

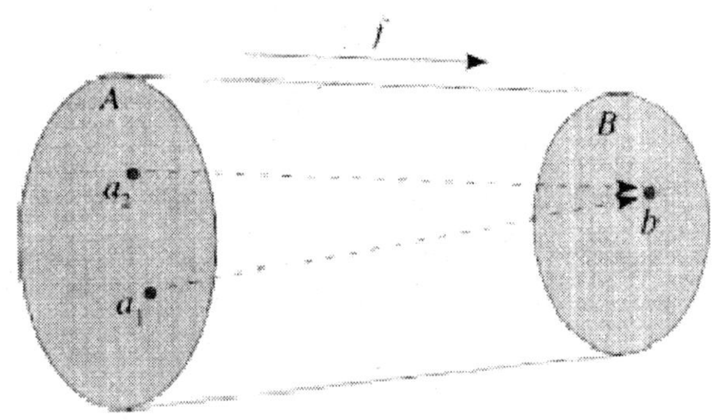

f is surjective iff Im f = B

Note. If $f: A \to B$, then $f(A)$ is sometimes called the range of f.

Examples 18.

A. $f: R \to R$; $f(x) = x^2 + 1$. Find $f(R)$ and $f[0,+\infty)$.

B. Identity maps are always surjective.

C. The inclusion : $C \to B$ is surjective iff $C = B$.

D. The canonical projections of a (possibly infinite) product.

E. Let S be any set and let $\approx$ be an equivalence relation on S. Denote the set of equivalence classes in S by $S/\approx$. Then there is a natural surjection $\nu: S \to S/\approx$.

Lemma 19. Let $f: A \to B$. Then:

(a) $f^1(f(C)) \supset C$ for all $C \subset A$ with equality iff f is injective.

(b) $f(f^1(D)) \subset D$ for a!! $D \subset$ B with equality iff f is surjective.

Proof. Exercise Set 2. We'll prove (a) in class.

Definition 20. $f : A \to B$ is bijective if it is both injective and surjective.

Examples 21.

A. Exponential map $R \to R^+$

B. Square root function

C. Inverse Trig functions

D. Multiplication by a non-zero real number

E. The identity map on any set

Definition 22. If $f : A \to B$ and $g : B \to C$, then their composite, $g \circ f : A \to C$, is the function specified by $g \circ f\,(a) = g(f(a))$.

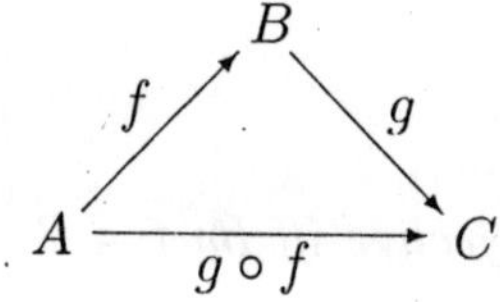

Lemma 23. Let $f : A \to B$ and $g : B \to C$. Then

(a) If f and g are injective, then so is $g \circ f$.

(b) If f and g are surjective, then so is $g \circ f$.

(c) If $g \circ f$ is injective, then so is f.

(d) If $g \circ f$ is surjective, then so is g.

Definition 24. $f : A \to B$ and $g : B \to A$ are called inverse maps if $g \circ f = 1_A$ and $f \circ g = 1_B$. In this event, we write $g = f^{1}$ (and say that g is the inverse of f) and $f = g^{-1}$. If f has an inverse, we say that f is invertible.

Theorem 25. (Inverse of a Function).

(a) $f : A \to B$ is invertible *iff* f is bijective

(b) The inverse of an invertible map is unique.

We prove (a) in class and leave (b) as an exercise.

Mathematical Induction and Properties of the Integers

The Axiom of Mathematical Induction is one of the central axioms of arithmetic. Here is one of the forms it can take:

Axiom of Mathematical Induction

If S is any subset of N such that:

(a) $0 \in S$;

(b) $n \in S \Rightarrow n + 1 \in S$,

then $S = N$.

The following is a theorem in "meta-mathematics":

Theorem 25. (Principle of Mathematical Induction).

If $P(n)$ is any proposition about the natural number n such that:

(a) $P(0)$ is true;

(b) If $P(n)$ is true, then $P(n + 1)$ is true, then $P(n)$ is true for all $n \in N$.

Proof. Let $S = \{n \in N \mid P(n) \text{ is true}\} \ldots$

Corollary 26. (General Principle of Mathematical Induction).

If $P(n)$ is any proposition about the natural number $n \geq k$ such that:

(a) $P(k)$ is true;

(b) If $P(n)$ is true, then $P(n + 1)$ is true, then $P(n)$ is true for all $n \geq k$.

Proof. Let $Q(n)$ be the proposition "$P(n + k)$ is true," and apply the theorem to Q.

Examples 27. We prove the following by induction:

A. $1 + 2 + \cdots + n = n(n + 1)^2$ for all $n \geq 1$.

B. De Morgan's Law for finite unions: If A, B_i $(i = 1, 2, \ldots)$ are any sets, then:

C. Every polynomial over Z factors as a product of irreducible polynomials over Z.

Definition 27. The integer a divides the integer b if there exists an integer k such that $b = ak$. When this is the case, we write $a|b$.

Definition 28. The positive integer h is the greatest common divisor (gcd) or highest common factor (hcf) of a and b if

(a) h is a divisor of a and b; (we write $h|a$ and $h|b$)

(b) If d is a divisor of a and b, then d is a divisor of h. That is,

$$d|a \text{ and } d|b)\ d|h$$

We denote the hcf h of a and b by (a, b).

Note. The hcf is always positive, so that $(\pm a, \pm b) = (a, b)$.

Proposition 29. (Existence and Properties of the hcf).

If a and b are integers, not both 0, then (a, b) exists. Moreover, there exist integers m and n such that

$$ma + nb = (a, b)$$

Proof: Here is the very elegant proof in Herstein. Let

$$M = \{ma + nb \mid m, n \in Z\}$$

Then M certainly contains positive integers (since a and b are not both zero), so let h be the smallest positive element of M. We claim that h is the hcf, which not only proves existence, but also the property

$$h = ma + nb.$$

Indeed, we need to show properties (a) and (b) above.

Property (a). Let us prove that h is a divisor of a. If $a = 0$, then h is certainly a divisor of a. If $a \neq 0$, then by the division algorithm we can divide a by h to obtain

$$a = ph + r \text{ with } 0 = r < h$$

(that's even true is a is negative!) Hence

$$\begin{aligned} r &= a - ph \\ &= a - p(ma + nb) \\ &= (1 - pm)a + (-pn)b \in M. \end{aligned}$$

But now $r \in M$ is non-negative, and is smaller than the smallest positive element, which forces it to be zero. In other words, h is a divisor of a, as claimed.

Property (b). If d is a common divisor of a and b, we must show that d is also a divisor of h. But the hypothesis implies that

$$a = kd \text{ and } b = ld$$

for some k and l. Hence

$$\begin{aligned} h &= ma + nb \\ &= m(kd) + m(ld) = d(mk + ml), \end{aligned}$$

showing that h is divisible by d, as required.

*Definition 30.*The integers a and b are relatively prime if $(a, b) = 1$.

Corollary 31. (Relatively Prime Integers).

If a and b are relatively prime, then there exist integers m and n such that

$$ma + nb = 1.$$

Definition 31. The integer $p > 1$ is prime if its only divisors are ± 1 and v $\pm p$. It follows that p is prime iff $(n, p) = 1$ or p for every integer n.

Lemma 32.

If $(a, b) = 1$ and $a \mid bc$, then $a \mid c$.

Proof. Choose m and n such that $ma + nb = 1$. Multiplying by c gives

$$mac + nbc = c.$$

Since a divides both terms on the left (the second term by hypothesis), it must divide the term on the right.

Corollary 33.

(a) If p is prime, and $p \mid bc$, then $p \mid b$ or $p \mid c$. More generally,

(b) If p is prime, and p divides any product of integers, then it divides at least one of them.

Proof. For part (a), assume $p \mid bc$ but p - b. Then $(p, b) = 1$, so by the lemma, $p|c$. Part (b) follows by an inductive argument.

Groups

Definition 34. A binary operation on a set S is a map $* : S \times S \to S$. In other words, the operation $*$ assigns to each pair (s, t) of elements in S an element $*(s, t)$, which we shall write as $s * t$, of S.

Examples 35.

A. Addition, subtraction and multiplication on R.

B. Division and multiplication in Q.

C. Composition in Map(X,X), the set of maps $X \to X$.

D. Multiplication of $n \times n$ matrices.

E. Concatenation of strings in a given set of symbols.

F. N is not closed under subtraction; hence subtraction is not a binary operation on N.

G. The quotient a/b is not defined for every pair of real numbers (a, b). Hence division is not a binary operation on R.

Recall that Z/nZ is the set of equivalence classes of integers modulo n.

Lemma 36. (Addition modula n).

The operation +: $Z/nZ \times Z/nZ \to Z/nZ$ given by $[m] + [n] = [m + n]$ is a well defined binary operation.

Some of these operations have the nice properties we ascribe to a group:

Definition 37. A group $(G,*)$ is a set G together with a binary operation $*$ on G such that:

(a) is associative.

(b) $(a * b) * c = a * (b * c)$ for all $a, b, c \in G$.

(c) There is an element $e \in G$ called an identity element such that $e * g = g * e = g$ for every $g \in G$.

(d) For every $g \in G$, there exists an element $g' \in G$, called an inverse of g such that $g * g' = g' * g = e$.

Examples 38.

A. $(Z, +)$

B. $(Q^*, \times)$

C. $(M(m, n), +)$

D. $(GL(n; Q), \times)$, $(GL(n; R), \times)$, and $(GL(n; C), \times)$

E. The set $C_n = \{\omega^0, \omega, \omega^2, \ldots, \omega^{n-1}\} \subset C$ of nth roots of unity, under multiplication.

F. $(Z/nZ, +)$ Compare its group structure with that of C_n.

G. The set of all invertible maps $R \to R$ under composition

H. The set S_A of all bijections of a set A, under composition

I. The unit circle, $S^1 = \{z \in C \mid |z| = 1\}$ under multiplication

J. Z is not a group under subtraction.

K. Z is not a group under multiplication.

L. Q is not a group under multiplication

Note: When we do not want to be explicit about the group operation, we shall leave it out, and write $a * b$ as ab or sometimes as $a + b$ when appropriate. Similarly, we shall write a group as G rather than $(G, *)$ when the group operation is understood.

Lemma 39. (Cancellation Law, Uniqueness of Identity and Inverses).

If G is a group, then the following are true:

(a) Left Cancellation: If g, h and k have the property that $gh = gk$, then $h = k$.

(b) Right Cancellation: If g, h and k have the property that $hg = kg$, then $h = k$.

(c) Uniqueness of the Identity: If e and e' are both identities in G, then $e = e'$.

(d) Uniqueness of Inverses: If g' and g'' are both inverses of $g \in G$, then $g' = g''$.

Note. As a consequence of part (d) of the lemma, we shall speak of the inverse of g and write it as g^{-1}. Similarly, we shall speak of the identity element of a group.

Lemma 40. (Product of Inverses).

If G is a group and $a, b \in G$, then $(ab)^{-1} = b^{-1}a^{-1}$.

The proof is in the exercise set.

Definition 41. An abelian group is a group G satisfying the commutative law:

$$ab = ba \text{ for all } a, b \in G.$$

Examples 42. Spot which of the above examples of groups are abelian.

A. The Dihedral Groups D_n

D_n is the set of symmetries of the regular n-gon (n rotations and n reflections). Illustrated in class. If a is rotation through

$2\pi/n$, and if b is reflection in the x-axis, then we can write D_n as:

$$D_n = \{e, a, a^2, \ldots, a^{n-1}, b, ba, ba^2, \ldots, ba^{n-1}\}$$

Checking that this indeed gives us a group is tedious, but we shall see in the exercises how to avoid this by realizing D_n as a group of 2×2 matrices. We can multiply elements according to the rule: $ba = a^{-1}b$. In terms of "generators and relations" we can also describe D_n as:

$$D_n = \langle a, b \mid a^n = b^2 = e, bab^{-1} = a^{-1} \rangle$$

B. The Symmetric Groups S_n

S_n is defined to be the group of permutations (self-bijections) of the set $\{1, 2, 3, \ldots, n\}$. In other words, $S_n = S\{_{1,2,3,\ldots,n}\}$. We look at the examples S_2 and S_3 and their multiplicative structure in class.

C. *The Quaternion Group Q_8*

Q_8 is defined by

$$Q_8 = \left\{ \begin{bmatrix} 1 & 0 \\ 0 & 1 \end{bmatrix}, \begin{bmatrix} -1 & 0 \\ 0 & -1 \end{bmatrix}, \begin{bmatrix} 0 & 1 \\ -1 & 0 \end{bmatrix}, \begin{bmatrix} 0 & -1 \\ 1 & 0 \end{bmatrix}, \right.$$

$$\left. \begin{bmatrix} 0 & i \\ i & 0 \end{bmatrix}, \begin{bmatrix} 0 & -i \\ -i & 0 \end{bmatrix}, \begin{bmatrix} i & 0 \\ 0 & -i \end{bmatrix}, \begin{bmatrix} -i & 0 \\ 0 & i \end{bmatrix} \right\}$$

The group operation is matrix multiplication, and i is the familiar complex number. We abbreviate these elements as follows:

$$Q_8 = \{1, -1, i, -i, j, -j, k, -k\}$$

In class, we verify the following:

$i^2 = j^2 = k^2 = -1$, and $ij = k = -ji$, $jk = i = -kj$, $ki = j = -ik$

Subgroups

First, some comments and definitions.

Remark. The notation + for the group operation will only be used for abelian groups, and then only for groups in which it stands for what is commonly thought of as the sum. Such a group will be called additive and its identity element will be written as 0 rather than e.

Definitions 43.

(1) The order of the finite group G is the cardinality of G as a set, and will be denoted by $|G|$.

(2) If S is a subset of G such that for every s and $t \in S$, one has $st \in S$, then we shall say that S is closed under the group operation in G. Note that S then inherits a binary operation from G. If S is a subset of G which is closed under the group operation, we refer to the resulting binary operation on S as induced operation on S.

Examples 44.

A. $2Z \subset Z$ is closed under the group operation (+) of Z.

B. $N \subset Z$ is also closed under addition.

C. The set of odd numbers is not a closed subset of Z under addition.

Definition 45. A subgroup of G is a subset $H \subset G$ such that:

(a) H is closed under the group operation;

(b) H is a group in its own right under the induced operation.

If H is a subgroup of G, we shall write $H < G$.

Examples 46.

A. Every group is a subgroup of itself.

B. The subset $\{e\} \subset G$ is a subgroup.

C. $2Z \subset Z$ is a subgroup because it is a group in its own right.

Note. A subgroup H of G must be a non-empty subset of G, since being a group in its own right implies that it contains the identity. In practice, the following criterion is extremely useful in checking that a given subset of G is a subgroup:

Proposition 47. (Test for a Subgroup). A subset $H \subset G$ is a subgroup of G iff:

(a) H is non empty and closed under the group operation

(b) H is closed under inverses: if $h \in H$, then $h^{-1} \in H$.

In words, this tells us that for a subset to be a subgroup, it must be nonempty and contain the products and inverses of all its elements. It also gives us the following mechanical test: How to check that H is a subgroup of G

(a) Show that $H \neq \theta$ and if a and $b \in H$, then $ab \in H$

(b) Show that, if $h \in H$, then $h^{-1} \in H$.

We can now give many more examples of subgroups.

Examples 48.

A. The subgroup $SL(n)$ of $GL(n)$ consisting of $n \times n$ matrices with determinant +1.

B. $G = GL(n)$; H the set of upper triangular matrices

C. Find all the subgroups of C_6.

D. Find all the subgroups of D_2.

E. Some finite subgroups of the circle group.

F. Subgroups of Z.

G. The continuous functions in $(R^R, +)$

Definition 49. A group G is called cyclic if it contains an element g such that every element of G is a (possibly negative) power of g. We refer to g as a generator of G.

Examples 50.

A. Generators of C_n; for $g = \omega^r$ to be a generator, some power of it must be ω. This translates to an interesting requirement on r.

B. Generators of Z.

C. Generators of mZ,

D. D_n is not cyclic.

E. Q is not cyclic under $\times$.

Definition 51. A proper subgroup of G is a subgroup $H < G$ such that $H \neq G$.

Definition and Proposition 52. If $g \in G$, let hgi denote the subset $\{g^n \mid n \in Z\}$. Then $\langle g \rangle$ is a subgroup of G, called the cyclic subgroup generated by g.

Examples 51.

A. Look at various such subgroups of C_n

B. $\langle a \rangle$ and $\langle b \rangle$ are subgroups of D_n.

C. The subgroup of $GL(n)$ generated by a rotation matrix.

D. Cyclic subgroups of Z.

Theorem 52 (Subgroups of Cyclic Groups).

Every subgroup of a cyclic group is cyclic.

Proof. Let G be a cyclic group, so that $G = \langle g \rangle$, say, and let $H < G$. Then H is a set of powers of g. Choose n to be the smallest positive exponent of elements in H;

$$n = \min\{i \in N \mid i > 0 \text{ and } g^i \in H\}.$$

Then we claim that every element of H is a power of $a = g^n$, giving the result. Indeed, if $h \in H$ is not the identity, then either h or h^{-1} is of the form g^m with $m > 0$, so that $m = n$. Dividing m by n gives

$$m = qn + r$$

with $r < n$ or $r = 0$, whence

$$g^m = (g^n)^q g^r, \text{ giving}$$
$$g^r = g^m (g^{-n})^q,$$

a product of elements of H, showing that $g^r \in H$. By the choice of m, we must have $r = 0$, giving

$$h \text{ (or } h^{-1}) = g^m = (g^n)^q,$$

proving the result.

Corollary 53. (Classification of subgroups of Z).

Every subgroup of Z is cyclic and of the form nZ.

The Permutation Groups

Recall that S^n is the group of all permutations of the set $\{1, 2, \ldots, n\}$.

Definition 54. The cycle $(n_1, n_2, \ldots, n_m)$ of distinct integers $n_i \leq n$ is the element σ of S_n defined by:

$$\sigma(j) \begin{cases} n_{i+1} & \text{if } j = n_i \text{ and } i \leq m-1 \\ n_1 & \text{if } j = n_m \\ j & \text{otherwise} \end{cases}$$

We call m the length of the cycle.

Examples 54. In class; including 1-cycles, 2-cycles, etc.

(a) $(n_1, n_2, \ldots, n_m) = (n_m, n_1, n_2, \ldots, n_{m-1}) = \ldots$

(b) $(n_1, n_2, \ldots, n_m)^{-1} = (n_m, n_{m-1}, \ldots, n_1)$

Multiplying Cycles Practice examples in class, and specifically (1, 2, 3, 4) (4, 5, 6, 7)

Examples 55. Some Permutation Groups

A. S_3

B. S_4

C. D_n for various n may be represented as a group of permutations-namely, what the rotations and reflections do to the vertices

D. Rigid motions of the cube.

Definition 56. The two cycles $(n_1, n_2, \ldots, n_m)$ and $(r_1, r_2, \ldots, r_s)$ are disjoint if $\{n_1, n_2, \ldots, n_m\}$ and $\{r_1, r_2, \ldots, r_s\}$ are disjoint as sets.

Note. Multiplication of disjoint cycles is commutative.

If $\sigma \in S_n$, define an equivalence relation on $\{1, 2, \ldots, n\}$ by $a \approx b$ if there exists an $r \geq 0$ with $\sigma^r(a) = b$.

Definition 57. The equivalence classes of the relation defined above are called the orbits in $\{1, 2, \ldots, n\}$ under σ.

Lemma 58. Each orbit in $\{1, 2, \ldots, n\}$ under $\sigma \in S_n$ has the form

$$\{k, \sigma(k), \sigma^2(k), \ldots, \sigma^{j-1}(k)\}$$

for some $k \in \{1, 2, \ldots, n\}$.

Proof. Let O be any orbit under σ, and choose $k \in O$. Consider the subset $S = \{k, \sigma(k), \sigma^2(k), \ldots\} \subset \{1, 2, \ldots, n\}$. By definition of the equivalence relation, every element of the orbit containing k must be a power of σ applied to k, and hence in S. Conversely, every element of S must be in the same orbit as k. It follows that the orbit of O of k is exactly the set S. That is,

$$O = S.$$

What, exactly, does S look like? Since S cannot have more than n elements, we must have repetitions; that is, $\sigma^i(k) = \sigma^j(k)$ for some $i < j < n$. Let j be the smallest integer ≥ 1 such that $\sigma^i(k) = \sigma^j(k)$ for some $i < j$. Claim: $i = 0$; that is, $\sigma^j(k) = 1$. Indeed, if $i > 0$, (so that $j \geq 2$) then apply σ^{-1} to both sides of the equation $\sigma^i(k) = \sigma^j(k)$, getting $\sigma^{i-1}(k) = \sigma^{j-1}(k)$, with j-1 ≥ 1, contradicting the fact that j was the smallest such integer. Because of the claim, it also follows that $k, \sigma(k), \sigma^2(k), \ldots, \sigma^{j-1}(k)$ are all distinct. Hence, O is the set described.

Theorem 59. (Disjoint Cycle Representation).

Every permutation of n letters is a product of disjoint cycles. Further, any two decompositions of into such a product give the same disjoint cycles.

Proof. If σ is a permutation of $N = \{1, 2, \ldots, n\}$, then N is a disjoint union of equivalence classes. Each of these classes is of the form $\{k, \sigma(k), \sigma^2(k), \ldots, \sigma^{j-1}(k)\}$ by the lemma. Let the disjoint equivalence classes be $\{k_1, \sigma(k_1), \sigma^2(k_1), \ldots, \sigma^{r1}(k_1)\}$, $\{k_2, \sigma(k_2), \sigma^2(k_2), \ldots, \sigma^{r2}(k_2)\}, \ldots, \{k_q, (k_q), \sigma(k_q), \ldots, \sigma^{rq}(k_q)\}$. We now observe that

$$\sigma = (k_1, \sigma(k_1), \sigma^2(k_1), \ldots, \sigma^{r1}(k_1))(k_2, (k_2), \sigma^2(k_2), \ldots, \sigma^{r2}(k_2)) \ldots (k_q, \sigma(k_q), \sigma^2(k_q), \ldots, \sigma^{rq}(k_q)),$$

a product of disjoint cycles.

Examples 60. Express the following as products of disjoint cycles:

A. (1, 2, 3)(3, 4, 5)

B. (1, 2)(2, 1, 3)(4, 3, 1)

C. (1, 2, 3)-1(3, 5, 6)(2, 1)

D. $(a, b)(b, c)(c, d)(e, f)(f, g)$

Definition 61. A cycle of length 2 is called a transposition.

Theorem 62. (Everything is 2-Cycles).

Every permutation of $\{1, 2, \ldots, n\}$ is a product of (not necessarily disjoint) transpositions.

Note. We can ignore cycles of length 1 in such a decomposition, or we can write, for example, (1) = (1, 2)(1, 2).

Lemma 63.(Effect of a 2-Cycle on the Number of Orbits).

If τ is a transposition and σ is any permutation in S_n, then the number of disjoint orbits in $\tau\sigma$ and σ differ by 1. (Note that disjoint orbits include those of length 1.)

Proof. Let $\tau = (a, b)$. We look at two cases:

Case 1. a is in one of the disjoint cycles of σ and b is in another. Then we see in class how to glue τ to the three cycles involved to get a single disjoint one. In other words, multiplication by τ has decreased the number of disjoint cycles by 1.

Case 2. a and b are in the same cycle of σ.

Then we observe that

$$(a, b)(a, x, y, z, \ldots, q, b, r, s, t, \ldots, k) = (a, x, y, z, \ldots, q)(b, r, s, t, \ldots, k).$$

In other words, it breaks up that cycle into two disjoint ones, thereby increasing the number of disjoint cycles by 1.

Theorem 64. (Parity is Well-Defined).

Any two decompositions of σ as a product of transpositions have the same parity. (That is, the number of transpositions is either odd for both or even for both.)

Proof. Suppose that the permutation $\sigma \in S_n$ can be expressed in two ways as product of 2-cycles:

$$\sigma = \tau_1 \tau_2 \ldots \tau_s = \epsilon_1 \epsilon_2 \ldots \epsilon t.$$

We must show that $s \equiv t \bmod 2$. But

$$\sigma = \tau_1 \tau_2 \ldots \tau_s \iota$$

where ι is the identity; $\iota = (1)(2)\ldots(n)$. By the lemma, the number of disjoint cycles in σ must therefore be congruent to $n + s$

mod 2. Similarly, applying this fact to the second decomposition gives

$$n + s \equiv n + t \bmod 2,$$

whence $s \equiv t \bmod 2$, as required.

Definition 64. A permutation that can be expressed as an even number of transpositions is called an even permutation. Otherwise, it is called an odd permutation. We write

$$sgn(\sigma) = \begin{cases} 1 & \text{if } \sigma \text{ is even,} \\ -1 & \text{if } \sigma \text{ is odd.} \end{cases}$$

Definition and Proposition 65

(a) The set of even permutations in S_n is a subgroup, called A_n, the alternating group on n letters.

(b) $|A_n| = \frac{|S_n|}{2}$

Cosets and Lagrange's Theorem

We now define a most peculiar relation on the elements of a group, arising from a given subgroup:

Definition 66. If G is a group and $H \subset G$ is a subgroup, define a relation on G by

$$a \approx H\ b \Leftrightarrow a^{-1}b \in H$$

We say that a is congruent to b mod H.

Now, saying that $a^{-1}b \in H$ is the same as saying that $a^{-1}b = h$ for some $h \in H$. In other words, $b = ah$ for some $h \in H$. Thus:

$$a \approx H\ b \Leftrightarrow a^{-1}b \in H$$

$$\Leftrightarrow b = ah \text{ for some } h \in H$$

Lemma 67. Congruence mod H is an equivalence relation. What do the equivalence classes look like?

Definition 68. If $H \subset G$ and $g \in G$, then define the left coset of H containing g as

$$gH = \{gh \mid h \in H\}$$

Proposition 68. (Congruence mod H and Cosets).

(a) The following are equivalent:

(i) $a \approx H\ b$

(ii) $a^{-1}b \in H$

(iii) $b = ah$ for some $h \in H$

(iv) $b \in aH$

(v) $bH \subset aH$

(vi) $bH = aH$

(b) The left cosets aH are the equivalence classes of equivalence mod H. Thus, two left cosets are either equal or disjoint (this being true of equivalence classes in general).

Proof. We prove (*a*) in class. For part (*b*), denote the equivalence class of $a \in G$ by $[a]$. One has

$b \in [a] \Leftrightarrow a \approx H_b$	Definition of equivalence classes
$\Leftrightarrow b \in aH$	By part (*a*), (*i*) $\Rightarrow$ (*iv*)

whence $[a] = aH$. That is, the equivalence classes are just the left cosets, as required.

2

Molecular Symmetry and Chemical Bonding

Group theory provides a very powerful set of tools that have found applications in many areas of chemistry and physics. It can be used to predict whether a given molecule will be chiral or polar, what type of vibrations it can undergo, whether it can absorb light of a particular polarisation, and which spectroscopic transitions will be excited if it can. Of primary interest to this course, it can be used to examine chemical bonding and to construct molecular orbitals. We will begin by learning how to classify molecules according to their symmetry.

Symmetry Operations and Symmetry Elements

A symmetry operation is an action that leaves an object looking the same after it has been carried out. For example, if we take a molecule of water and rotate it by 180° an axis passing through the central *O* atom (between the two *H* atoms), it will look the same as before. It will also look the same if we reflect it through either of two mirror planes, as shown in the figure below.

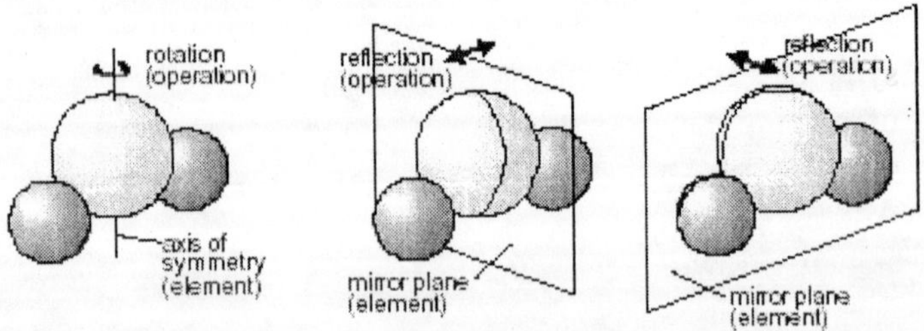

Each symmetry operation has a corresponding symmetry element, which is the axis, plane, line or point with respect to which the symmetry operation is carried out. The symmetry element consists of all the points that stay in the same place when the symmetry operation is performed. In a rotation, the line of points that stay in the same place constitute a symmetry axis; in a reflection the points that remain unchanged make up a plane of symmetry. The symmetry elements that a molecule may possess are:

1. E – the identity. The identity operation consists of doing nothing, and the corresponding symmetry element is the entire molecule. Every molecule has at least this element.
2. C_n – an n-fold axis of rotation. Rotation by 360°/n leaves the molecule unchanged. The H_2O molecule above has a C_2 axis. Some molecules have more than one C_n axis, in which case the one with the highest value of n is called the principal axis. Note that by convention rotations are counter-clockwise about the axis.
3. σ - a plane of symmetry. Reflection in the plane leaves the molecule looking the same. In a molecule that also has an axis of symmetry, a mirror plane that includes the axis is called a vertical mirror plane and is labelled σ_v, while one perpendicular to the axis is called a horizontal mirror plane and is labelled σ_h. A vertical mirror plane that bisects the angle between two C_2 axes is called a dihedral mirror plane, σ_d.

4. i – a centre of symmetry. Inversion through the centre of symmetry leaves the molecule unchanged. Inversion consists of passing each point through the centre of inversion and out to the same distance on the other side of the molecule. An example of a molecule with a centre of inversion is shown below.

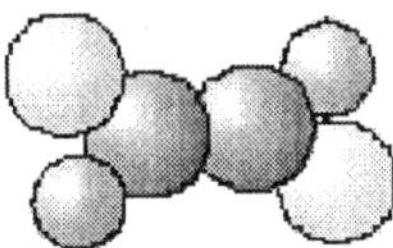

5. S_n – an n-fold improper rotation axis (also called a rotary-reflection axis). The rotary reflection operation consists of rotating through an angle $360°/n$ about the axis, followed by reflecting in a plane perpendicular to the axis. Note that S_1 is the same as reflection and S_2 is the same as inversion. The molecule shown above has two S_2 axes.

The identity E and rotations C_n are symmetry operations that could actually be carried out on a molecule. For this reason they are called proper symmetry operations. Reflections, inversions and improper rotations can only be imagined (it is not actually possible to turn a molecule into its mirror image or inversion without some fairly drastic rearrangement of chemical bonds) and as such, are termed improper symmetry operations.

Symmetry Classification of Molecules – Point Groups

There are only certain combinations of symmetry elements that can be present in a molecule (or any other object). As a result, we can group together molecules that possess the same symmetry elements and classify molecules according to their symmetry. These groups of symmetry elements are called point groups (due to the fact that there is at least one point in space that remains unchanged no matter which symmetry operation from the group is applied). There are two systems of notation for labelling symmetry groups, called the Schoenflies and Hermann- Mauguin (or International) systems. The symmetry of individual molecules is usually described using the Schoenflies notation, and we shall be using this notation for the remainder of the course. The molecular point groups are listed below.

1. C_1 – contains only the identity (a C_1 rotation is a rotation by 360° and is the same as the identity operation E) e.g. CHDFCl.

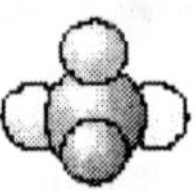

2. C_i– contains the identity E and a centre of inversion i.

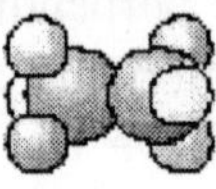

3. C_s – contains the identity E and a plane of reflection σ.

4. C_n – contains the identity and an n-fold axis of rotation.

5. C_{nv} – contains the identity, an n-fold axis of rotation, and n vertical mirror planes σ_v.

6. C_{nh} – contains the identity, an n-fold axis of rotation, and a horizontal reflection plane σ_h (note that in C_{2h} this

combination of symmetry elements automatically implies a centre of inversion).

7. D_n – contains the identity, an n-fold axis of rotation, and n 2-fold rotations about axes perpendicular to the principal axis.

8. D_{nh} – contains the same symmetry elements as D_n with the addition of a horizontal mirror plane.

9. D_{nd} – contains the same symmetry elements as D_n with the addition of n dihedral mirror planes.

10. S_n – contains the identity and one S_n axis. Note that molecules only belong to S_n if they have not already been classified in terms of one of the preceding point groups (e.g. S_2 is the same as C_i, and a molecule with this symmetry would already have been classified).

The following groups are the cubic groups, which contain more than one principal axis. They separate into the tetrahedral groups (T_d, T_h and T) and the octahedral groups (O and O_h). The icosahedral group also exists but is not included below.

11. T_d – contains all the symmetry elements of a regular tetrahedron, including the identity, 4 C_3 axes, 3 C_2 axes, 6 dihedral mirror planes, and 3 S_4 axes e.g. CH_4.
12. T – as for T_d but no planes of reflection.
13. T_h – as for T but contains a centre of inversion.
14. O_h – the group of the regular octahedron e.g. SF_6.

15. O – as for O_h but with no planes of reflection.

The final group is the full rotation group R_3, which consists of an infinite number of C_n axes with all possible values of n and describes the symmetry of a sphere. Atoms (but no molecules) belong to R_3, and the group has important applications in atomic quantum mechanics. However, we won't be treating it any further here. Once you become more familiar with the symmetry elements and point groups described above, you will find it quite straightforward to classify a molecule in terms of its point group. In the meantime, the flowchart shown below provides a step-by-step approach to the problem.

Symmetry and Physical Properties

Carrying out a symmetry operation on a molecule must not change any of its physical properties. It turns out that this has some interesting consequences, allowing us to predict whether or not a molecule may be chiral or polar on the basis of its point group.

Polarity

For a molecule to have a permanent dipole moment, it must have an asymmetric charge distribution. The point group of the molecule not only determines whether the molecule may have a dipole moment, but also in which direction(s) it may point. If a molecule has a C_n axis with n>1, it cannot have a dipole moment perpendicular to the axis of rotation (for example, a C_2 rotation would interchange the ends of such a dipole moment and reverse the polarity, which is not allowed – rotations with higher values of n would also change the direction in which the dipole points). Any dipole must lie parallel to a C_n axis. Also, if the point group of the molecule contains any symmetry operation that would interchange the two ends of the molecule, such as a σ_h mirror plane or a C_2 rotation perpendicular to the principal axis, then there cannot be a dipole moment along the axis. The only groups compatible with a dipole moment are C_n, C_{nv} and C_s. In molecules belonging to C_n or C_{nv} the dipole must lie along the axis of rotation.

Chirality

We mentioned enantiomeric pairs of molecules. Such molecules are said to be chiral, meaning that they cannot be superimposed on their mirror image. Formally, the symmetry element that precludes a molecule from being chiral is a rotation-reflection axis S_n. Such an axis is often implied by other symmetry elements present in a group. For example, a point group that has C_n and σ_h as elements will also have S_n. Similarly, a centre of inversion is equivalent to S_2. As a rule of thumb, a molecule definitely cannot have be chiral if it has a centre of inversion or a mirror plane of any type (σ_h, σ_ϖ or σ_d), but if these symmetry elements are absent the molecule should be checked carefully for an S_n axis before it is assumed to be chiral.

Combining Symmetry Pperations: 'Group Multiplication'

Now we will investigate what happens when we apply two symmetry operations in sequence. As an example, consider the NH_3 molecule, which belongs to the C_{3v} point group. Consider what happens if we apply a C_3 rotation followed by a σ_v reflection. We write this combined operation $\sigma_v C_3$ (when written, symmetry operations operate on the thing directly to their right, just as operators do in quantum mechanics – we therefore have to work backwards from right to left from the notation to get the correct order in which the operators are applied). As we shall soon see, the order in which the operations are applied is important.

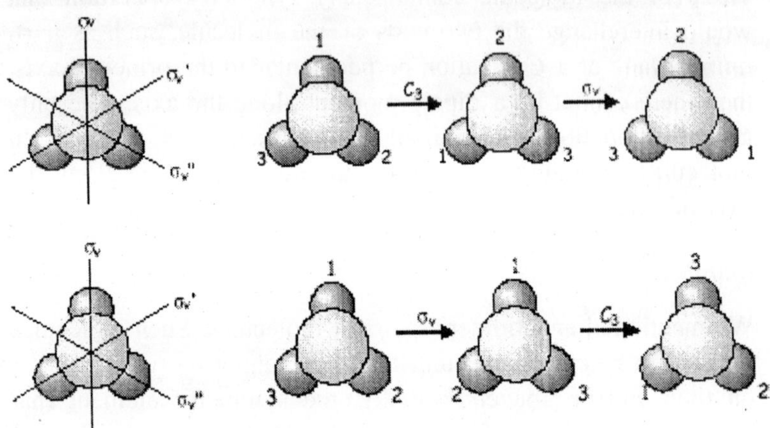

The combined operation $\sigma_v C_3$ is equivalent to σ_v', which is also a symmetry operation of the C_{3v} point group. Now let's see what happens if we apply the operators in the reverse order i.e. $C_3\sigma_v$. Again, the combined operation $C_3\sigma_v$ is equivalent to another operation of the point group, this time σ_v''. There are two important points that are illustrated by this example:

1. The order in which two operations are applied is important. For two symmetry operations *A* and *B*, *AB* is not necessarily the same as *BA*, i.e. symmetry operations do not in general commute. In some groups the symmetry elements do commute; such groups are said to be Abelian.

2. If two operations from the same point group are applied in sequence, the result will be equivalent to another operation from the point group. Symmetry operations that are related to each other by other symmetry operations of the group are said to belong to the same class. In NH_3, the three mirror

planes σ_v, σ_v' and σ_v'' belong to the same class (related to each other through a C_3 rotation), as do the rotations C_3^+ and C_3^- (anticlockwise and clockwise rotations about the principal axis, related to each other by a vertical mirror plane).

The effects of applying two symmetry operations in sequence within a given point group are summarised in group multiplication tables. As an example, the complete group multiplication table for C_{3v} is shown below. The operations written along the first row of the table are carried out first, followed by those written in the first column.

C_{3v}	E	C_3^+	C_3^-	σ_v	σ_v'	σ_v''
E	E	C_3^+	C_3^-	σ_v	σ_v'	σ_v''
C_3^+	C_3^+	C_3^-	E	σ_v'	σ_v''	σ_v
C_3^-	C_3^-	E	C_3^+	C_v''	σ_v	σ_v'
σ_v	σ_v	σ_v''	σ_v'	E	C_3^-	C_3^+
σ_v'	σ_v'	σ_v'	σ_v''	C_3^+	E	C_3^+
σ_v''	σ_v''	σ_v'	σ_v	C_3^-	C_3^+	E

Mathematical Definition of a Group

Now that we have explored some of the properties of symmetry operations and elements and their behaviour within point groups, we are ready to introduce the formal mathematical definition of a group. A mathematical group is defined as a set of elements (g_1, g_2, g_3...) together with a rule for forming combinations $g_i g_j$. The number of elements h is called the order of the group. For our purposes, the elements are the symmetry operations of a molecule and the rule for combining them is the sequential application of symmetry operations investigated. The elements of the group and the rule for combining them must satisfy the following criteria.

1. The group must include the identity E, for which $E_{g_i} = g_i E = g_i$ for all the elements of the group.
2. The elements must satisfy the group property that the combination of any pair of elements is also an element of the group.
3. Each element g_i must have an inverse g_i^{-1}, which is also an element of the group, such that $g_i g_i^{-1} = g_i^{-1} g_i = E$ (e.g. in C_{3v} the inverse of C_3^+ is C_3^-, the inverse of σ_v is σ_v; the inverse g_i^{-1} 'undoes' the effect of the symmetry operation g_i).
4. The rule of combination must be associative i.e. $g_i(g_j g_k) = (g_i g_j)g_k$

The above definition does not require the elements to commute (which would require $g_i g_k = g_k g_i$). As we discovered in the C_{3v} example above, in many groups the outcome of consecutive application of two symmetry operations depends on the order in which the operations are applied. Groups for which the elements do not commute are called non-Abelian groups; those for which the elements do commute are Abelian.

Group theory is an important area in mathematics, and luckily for chemists the mathematicians have already done most of the work for us. Along with the formal definition of a group comes a comprehensive mathematical framework that allows us to carry out a rigorous treatment of symmetry in molecular systems and learn about its consequences.

Many problems involving operators or operations (such as those found in quantum mechanics or group theory) may be reformulated in terms of matrices. Any of you who have come across transformation matrices before will know that symmetry operations such as rotations and reflections may be represented by matrices. It turns out that the set of matrices representing the symmetry operations in a group obey all the conditions laid out above in the mathematical definition of a group, and using matrix representations of symmetry operations simplifies carrying out calculations in group theory. Before we learn how to use matrices in group theory, it will probably be helpful to review some basic definitions and properties of matrices.

Matrices

An $n \times m$ matrix is a two dimensional array of numbers with n rows and m columns. The integers n and m are called the dimensions of the matrix. If $n = m$ then the matrix is square. The numbers in the matrix are known as matrix elements (or just elements) and are usually given subscripts to signify their position in the matrix e.g. an element a_{ij} would occupy the ith row and j^{th} column of the matrix.

$$\text{e.g. } M = \begin{pmatrix} 1 & 2 & 3 \\ 4 & 5 & 6 \\ 7 & 8 & 9 \end{pmatrix} \text{ is a } 3 \times 3$$

matrix with $a_{11} = 1$, $a_{12} = 2$, $a_{13} = 3$, $a_{21} = 4$ etc

In a square matrix, diagonal elements are those for which $i = j$ (the numbers 1, 5 and 9 in the above example). Off diagonal elements are those for which $i \neq j$ (2, 3, 4, 6, 7, and 8 in the above example). If all the off-diagonal elements are equal to zero then we have a diagonal matrix. We will see later that diagonal matrices are of considerable importance in group theory. A unit matrix or identity matrix (usually given the symbol I) is a diagonal matrix in which all the diagonal elements are equal to 1. A unit matrix acting on another matrix has no effect – it is the same as the identity operation in group theory and is analogous to multiplying a number by 1 in everyday arithmetic.

The transpose A^T of a matrix A is the matrix that results from interchanging all the rows and columns. A symmetric matrix is the same as its transpose ($A^T = A$ i.e. $a_{ij} = a_{ji}$ for all values of i and j). The transpose of matrix M above (which is not symmetric) is

$$M^T = \begin{pmatrix} 1 & 4 & 7 \\ 2 & 5 & 8 \\ 3 & 6 & 9 \end{pmatrix}$$

The sum of the diagonal elements in a square matrix is called the trace (or character) of the matrix (for the above matrix, the trace is $\chi = 1 + 5 + 9 = 15$). The traces of matrices representing symmetry operations will turn out to be of great importance in group theory. A vector is just a special case of a matrix in which one of the dimensions is equal to 1. An $n \times 1$ matrix is a column vector; a $1 \times m$ matrix is a row vector. The components of a vector are usually only labelled with one index. A unit vector has one element equal to 1 and the others equal to zero (it is the same as one row or column of an identity matrix). We can extend the idea further to say that a single number is a matrix (or vector) of dimension 1×1.

Matrix Algebra

i) Two matrices with the same dimensions may be added or subtracted by adding or subtracting the elements occupying the same position in each matrix.

$$\text{e.g.} \quad A = \begin{pmatrix} 1 & 0 & 2 \\ 2 & 2 & 1 \\ 3 & 2 & 0 \end{pmatrix} \quad B = \begin{pmatrix} 2 & 0 & -2 \\ 1 & 0 & 1 \\ 1 & -1 & 0 \end{pmatrix}$$

$$A + B = \begin{pmatrix} 3 & 0 & 0 \\ 3 & 2 & 2 \\ 4 & 1 & 0 \end{pmatrix} \quad A - B = \begin{pmatrix} -1 & 0 & 4 \\ 1 & 2 & 0 \\ 2 & 3 & 0 \end{pmatrix}$$

ii) A matrix may be multiplied by a constant by multiplying each element by the constant.

$$\text{e.g.} \quad 4B = \begin{pmatrix} 8 & 0 & -8 \\ 4 & 0 & 4 \\ 4 & -4 & 0 \end{pmatrix} \quad 3A = \begin{pmatrix} 3 & 0 & 6 \\ 6 & 6 & 3 \\ 9 & 6 & 0 \end{pmatrix}$$

iii) Two matrices may be multiplied together provided that the number of columns of the first matrix is the same as the number of rows of the second matrix i.e. an $n \times m$ matrix may be multiplied by an $m \times l$ matrix. The resulting matrix will have dimensions $n \times l$. To find the element a_{ij} in the product matrix, we take the dot product of row i of the first matrix and column j of the second matrix (i.e. we multiply consecutive elements together from row i of the first matrix and column j of the second matrix and add them together i.e. $c_{ij} = \Sigma_k a_{ik} b_{kj}$ e.g. in the 3×3 matrices A and B used in the above examples, the first element in the product matrix $C = AB$ is $c_{11} = a_{11}b_{11} + a_{12}b_{21} + a_{13}b_{31}$

$$AB = \begin{pmatrix} 1 & 0 & 2 \\ 2 & 2 & 1 \\ 3 & 2 & 0 \end{pmatrix} \begin{pmatrix} 2 & 0 & -2 \\ 1 & 0 & 1 \\ 1 & -1 & 0 \end{pmatrix} = \begin{pmatrix} 4 & -2 & -2 \\ 7 & -1 & -2 \\ 8 & 0 & -4 \end{pmatrix}$$

An example of a matrix multiplying a vector is

$$Av = \begin{pmatrix} 1 & 0 & 2 \\ 2 & 2 & 1 \\ 3 & 2 & 0 \end{pmatrix} \begin{pmatrix} 1 \\ 2 \\ 3 \end{pmatrix} = \begin{pmatrix} 7 \\ 9 \\ 7 \end{pmatrix}$$

Matrix multiplication is not generally commutative, a property that mirrors the behaviour found earlier for symmetry operations within a point group.

Inverse Matrices and Determinants

If two square matrices A and B multiply together to give the identity matrix I (i.e. AB = I) then B is said to be the inverse of A (written A^{-1}). If B is the inverse of A then A is also the inverse of B. Recall that one of the conditions imposed upon the symmetry operations in a group is that each operation must have an inverse. It follows by analogy that any matrices we use to represent symmetry elements must also have inverses. It turns out that a square matrix only has an inverse if its determinant is non-zero.

For this reason (and others which will become apparent later on when we need to solve equations involving matrices) we need to learn a little about matrix determinants and their properties. For every square matrix, there is a unique function of all the elements that yields a single number called the determinant. Initially it probably won't be particularly obvious why this number should be useful, but matrix determinants are of great importance both in pure mathematics and in a number of areas of science. Historically, determinants were actually around before matrices. They arose originally as a property of a system of linear equations that 'determined' whether the system had a unique solution. As we shall see later, when such a system of equations is recast as a matrix equation this property carries over into the matrix determinant.

There are two different definitions of a determinant, one geometric and one algebraic. In the geometric interpretation, we consider the numbers across each row of an $n \times n$ matrix as coordinates in n-dimensional space. In a one-dimensional matrix (i.e. a number), there is only one coordinate, and the determinant can be interpreted as the (signed) length of a vector from the origin to this point. For a 2×2 matrix we have two coordinates in a plane, and the determinant is the (signed) area of the parallelogram that includes these two points and the origin. For a 3×3 matrix the determinant is the (signed) volume of the parallelepiped that includes the three points (in three-dimensional space) defined by the matrix and the origin. This is illustrated on next page. The idea extends up to higher dimensions in a similar way. In some sense then, the determinant is therefore related to the size of a matrix.

The algebraic definition of the determinant of an $n \times n$ matrix is a sum over all the possible products of n elementstaken from different rows and columns. Each term in the sum is given a positive or a negative sign depending on whether the number of permutations required to put the indices of the elements in numerical order is even or odd (a permutation simply consists of swapping the positions of two numbers). The number of terms in the sum is the number of permutations of n numbers, which is generally equal to $n!$ (2 for a 2 × 2 matrix, 6 for a 3 × 3 matrix etc)

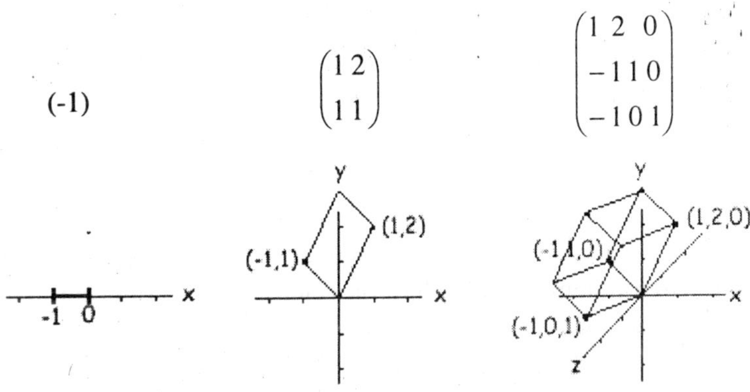

For example, for a two-dimensional matrix

$$\begin{pmatrix} a_{11} & a_{12} \\ a_{21} & a_{22} \end{pmatrix}$$

where the subscripts label the row and column positions of the elements, there are 2 possible products involving elements from different rows and column, $a_{11}a_{22}$ and $a_{12}a_{21}$. In the second term, the 2 and 1 of a_{21} would have to be swapped around to put the indices in numerical order (one permutation), so this term takes a negative sign, and the determinant is $a_{11}a_{22}$-$a_{12}a_{21}$.

For a 3 × 3 matrix

$$\begin{pmatrix} a_{11} & a_{12} & a_{13} \\ a_{21} & a_{22} & a_{23} \\ a_{31} & a_{32} & a_{33} \end{pmatrix}$$

the possible combinations of elements from different rows and columns, together with the sign from the number of permutations required to put their indices in numerical order are:

$a_{11}a_{22}a_{33}$

$-a_{11}a_{23}a_{32}$

$-a_{12}a_{21}a_{33}$

$-a_{12}a_{23}a_{31}$

$a_{13}a_{21}a_{32}$

$-a_{13}a_{22}a_{31}$

and the determinant is simply the sum of these terms.

This may all seem a little complicated, but in general there is a fairly systematic procedure for calculating determinants. The determinant of a matrix A is usually written det(A) or $|A|$.

For a 2×2 matrix

$$A = \begin{pmatrix} a & b \\ c & d \end{pmatrix} \qquad \det(A) = |A| = \begin{vmatrix} a & b \\ c & d \end{vmatrix} = ad-bc$$

For a 3×3 matrix

$$B = \begin{pmatrix} a & b & c \\ d & e & f \\ g & h & i \end{pmatrix} \qquad \det(B) = a\begin{vmatrix} e & f \\ h & i \end{vmatrix} - b\begin{vmatrix} d & f \\ g & i \end{vmatrix} + c\begin{vmatrix} d & e \\ g & h \end{vmatrix}$$

For a 4×4 matrix

$$C = \begin{pmatrix} a & b & c & d \\ e & f & g & h \\ i & j & k & l \\ m & n & o & p \end{pmatrix}$$

$$\det(C) = a\begin{vmatrix} f & g & h \\ j & k & l \\ n & o & p \end{vmatrix} - b\begin{vmatrix} e & g & h \\ i & k & l \\ m & o & p \end{vmatrix} + c\begin{vmatrix} e & f & h \\ i & j & l \\ m & n & p \end{vmatrix} - d\begin{vmatrix} e & f & g \\ i & j & k \\ m & n & o \end{vmatrix}$$

and so on in higher dimensions. Note that the submatrices in the 3 × 3 example above are just the matrices formed from the original matrix B that don't include any elements from the same row or column as the premultiplying factors from the first row.

Matrix determinants have a number of important properties:

(i) The determinant of the identity matrix is 1.

$$\text{e.g.} \quad \begin{vmatrix} 1 & 0 \\ 0 & 1 \end{vmatrix} = \begin{vmatrix} 1 & 0 & 0 \\ 0 & 1 & 0 \\ 0 & 0 & 1 \end{vmatrix} = 1$$

(ii) The determinant of a matrix is the same as the determinant of its transpose i.e. $\det(A) = \det(A^T)$

$$\text{e.g.} \quad \begin{vmatrix} a & b \\ c & d \end{vmatrix} = \begin{vmatrix} a & c \\ b & d \end{vmatrix}$$

(iii) The determinant changes sign when any two rows or any two columns are interchanged

$$\text{e.g.} \quad \begin{vmatrix} a & b \\ c & d \end{vmatrix} = -\begin{vmatrix} b & a \\ d & c \end{vmatrix} = -\begin{vmatrix} c & d \\ a & b \end{vmatrix} = \begin{vmatrix} d & c \\ b & a \end{vmatrix}$$

(iv) The determinant is zero if any row or column is entirely zero, or if any two rows or columns are equal or a multiple of one another.

$$\text{e.g.} \quad \begin{vmatrix} 1 & 2 \\ 0 & 0 \end{vmatrix} = 0, \quad \begin{vmatrix} 1 & 2 \\ 2 & 4 \end{vmatrix} = 0$$

(v) The determinant is unchanged by adding any linear combination of rows (or columns) to another row (or column).

(vi) The determinant of the product of two matrices is the same as the product of the determinants of the two matrices i.e. $\det(AB) = \det(A)\det(B)$.

The requirement that in order for a matrix to have an inverse it must have a non-zero determinant follows from property (vi). As mentioned previously, the product of a matrix and its inverse yields the identity matrix I. We therefore have:

$$\det(A^{-1}A) = \det(A^{-1})\det(A) = \det(I)$$
$$\det(A^{-1}) = \det(I)/\det(A) = 1/\det(A)$$

It follows that a matrix A can only have an inverse if its determinant is non-zero, otherwise the determinant of its inverse would be undefined.

Transformation Matrices

Matrices can be used to map one set of coordinates or functions onto another set. Matrices used for this purpose are called transformation matrices. In group theory, we can use transformation matrices to carry out the various symmetry operations considered at the beginning of the course. As a simple example, we will investigate the matrices we would use to carry out some of these symmetry operations on a vector (x, y).

The Identiy Operation

The identity operation leaves the vector unchanged, and as you may already suspect, the appropriate matrix is the identity matrix.

$$(x,y)\begin{pmatrix} 1 & 0 \\ 0 & 1 \end{pmatrix} = (x,y)$$

Reflection in a Plane

The simplest example of a reflection matrix corresponds to reflecting the vector (x, y) in either the x or y axes. Reflection in the x axis maps y to $-y$, while reflection in the y axis maps x to $-x$. The appropriate matrix is very like the identity matrix but with a change in sign for the appropriate element. Reflection in the x axis transforms the vector (x, y) to $(x, -y)$, and the appropriate matrix is

$$(x,y)\begin{pmatrix} 1 & 0 \\ 0 & -1 \end{pmatrix} = (x,-y)$$

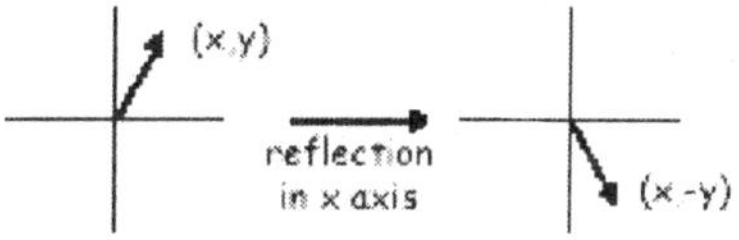

Reflection in the y axis transforms the vector (x, y) to $(-x, y)$, and the appropriate matrix is

$$(x,y)\begin{pmatrix} -1 & 0 \\ 0 & 1 \end{pmatrix} = (-x,y)$$

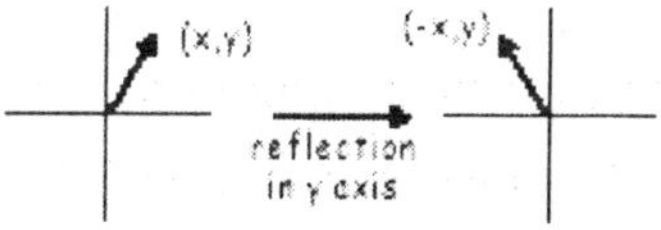

More generally, matrices can be used to represent reflections in any plane (or line in 2D). For example, reflection in the 45° axis shown below maps (x, y) onto $(-y, x)$.

$$(x,y)\begin{pmatrix} 0 & -1 \\ -1 & 0 \end{pmatrix} = (-y,-x)$$

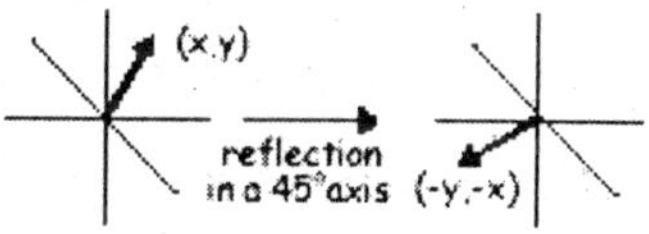

Rotation about an Axis

In two dimensions, the appropriate matrix to represent rotation by an angle θ about the origin is

$$R(\theta) = \begin{pmatrix} \cos\theta & -\sin\theta \\ \sin\theta & \cos\theta \end{pmatrix}$$

In three dimensions, rotations about the x, y and z axes acting on a vector (x, y, z) are represented by the following matrices.

$$R_x(\theta) = \begin{pmatrix} 1 & 0 & 0 \\ 0 & \cos\theta & -\sin\theta \\ 0 & \sin\theta & \cos\theta \end{pmatrix}$$

$$R_y(\theta) = \begin{pmatrix} \cos\theta & 0 & -\sin\theta \\ 0 & 1 & 0 \\ \sin\theta & 0 & \cos\theta \end{pmatrix} \qquad R_z(\theta) = \begin{pmatrix} \cos\theta & -\sin\theta & 0 \\ \sin\theta & \cos\theta & 0 \\ 0 & 0 & 1 \end{pmatrix}$$

Matrix Representations of Groups

We are now ready to integrate what we have just learned about matrices with group theory. The symmetry operations in a group may be represented by a set of transformation matrices $\Gamma(g)$, one for each symmetry element g. Each individual matrix is called a representative of the corresponding symmetry operation, and the complete set of matrices is called a matrix representation of the group. The matrix representatives act on some chosen basis set of functions, and the actual matrices making up a given representation will depend on the basis that has been chosen. The representation is then said to span the chosen basis. In the examples above we were looking at the effect of some simple transformation matrices on an arbitrary vector (x, y). The basis was therefore a pair of unit vectors pointing in the x and y directions. In most of the examples we will be considering in this course, we will use sets of atomic orbitals as basis functions for matrix representations.

Before proceeding any further, we must check that a matrix representation of a group obeys all of the rules set out in the formal mathematical definition of a group.

1. The first rule is that the group must include the identity operation E (the 'do nothing' operation). We showed above that the matrix representative of the identity operation is simply the identity matrix. As a consequence, every matrix representation includes the appropriate identity matrix.
2. The second rule is that the combination of any pair of elements must also be an element of the group (the group property). If we multiply together any two matrix representatives, we should get a new matrix which is a representative of another symmetry operation of the group. In fact, matrix representatives multiply together to give new representatives in exactly the same way as symmetry operations combine according to the group multiplication table. For example, in the C_{3v} point group, we showed that the combined symmetry operation $C_3\sigma_v$ is equivalent to σ_v''. In a matrix representation of the group, if the matrix representatives of C_3 and σ_v are multiplied together, the result will be the representative of σ_v''.
3. The third rule states that every operation must have an inverse, which is also a member of the group. The combined effect of carrying out an operation and its inverse is the same as the identity operation. It is fairly easy to show that matrix representatives satisfy this criterion. For example, the inverse of a reflection is another reflection, identical to the first. In matrix terms we would therefore expect that a reflection matrix was its own inverse, and that two identical reflection matrices multiplied together would give the identity matrix. This turns out to be true, and can be verified using any of the reflection matrices in the examples above. The inverse of a rotation matrix is another rotation matrix corresponding to a rotation of the opposite sense to the first.
4. The final rule states that the rule of combination of symmetry elements in a group must be associative. This is automatically satisfied by the rules of matrix multiplication.

Example: A matrix representation of the C_{3v} point group (the ammonia molecule).

The first thing we need to do before we can construct a matrix representation is to choose a basis. For NH_3, we will select a basis (s_N, s_1, s_2, s_3) that consists of the valence s orbitals on the nitrogen and the three hydrogen atoms. We need to consider what happens to this basis when it is acted on by each of the symmetry operations in the C_{3v} point group, and determine the matrices that would be required to produce the same effect. The basis set and the symmetry operations in the C_{3v} point group are summarised in the figure below.

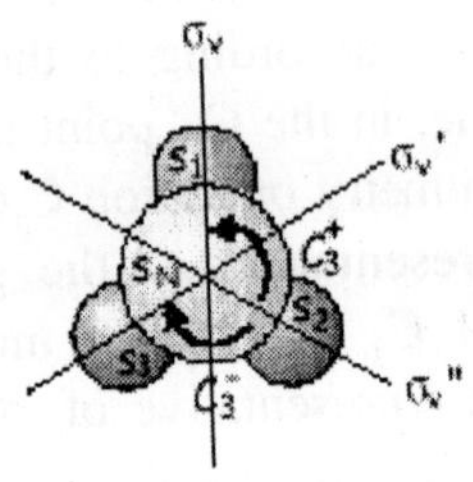

The effects of the symmetry operations on chosen basis are as follows:

$$
\begin{array}{ll}
E & (s_N, s_1, s_2, s_3) \rightarrow (s_N, s_1, s_2, s_3) \\
C_3^+ & (s_N, s_1, s_2, s_3) \rightarrow (s_N, s_3, s_1, s_2) \\
C_3^- & (s_N, s_1, s_2, s_3) \rightarrow (s_N, s_2, s_3, s_1) \\
\sigma_v & (s_N, s_1, s_2, s_3) \rightarrow (s_N, s_1, s_3, s_2) \\
\sigma_{v'} & (s_N, s_1, s_2, s_3) \rightarrow (s_N, s_2, s_1, s_3) \\
\sigma_v'' & (s_N, s_1, s_2, s_3) \rightarrow (s_N, s_3, s_2, s_1)
\end{array}
$$

By inspection, the matrices that carry out the same transformations are:

$$\Gamma(E) \quad (s_N, s_1, s_2, s_3)\begin{pmatrix} 1 & 0 & 0 & 0 \\ 0 & 1 & 0 & 0 \\ 0 & 0 & 1 & 0 \\ 0 & 0 & 0 & 1 \end{pmatrix} = (s_N, s_1, s_2, s_3)$$

$$\Gamma(C_3^+) \quad (s_N, s_1, s_2, s_3)\begin{pmatrix} 1 & 0 & 0 & 0 \\ 0 & 0 & 1 & 0 \\ 0 & 0 & 0 & 1 \\ 0 & 1 & 0 & 0 \end{pmatrix} = (s_N, s_3, s_1, s_2)$$

$$\Gamma(C_3^-) \quad (s_N, s_1, s_2, s_3)\begin{pmatrix} 1 & 0 & 0 & 0 \\ 0 & 0 & 0 & 1 \\ 0 & 1 & 0 & 0 \\ 0 & 0 & 1 & 0 \end{pmatrix} = (s_N, s_2, s_3, s_1)$$

$$\Gamma(\sigma_v) \quad (s_N, s_1, s_2, s_3)\begin{pmatrix} 1 & 0 & 0 & 0 \\ 0 & 1 & 0 & 0 \\ 0 & 0 & 0 & 1 \\ 0 & 0 & 1 & 0 \end{pmatrix} = (s_N, s_1, s_3, s_2)$$

$$\Gamma(\sigma_v') \quad (s_N, s_1, s_2, s_3)\begin{pmatrix} 1 & 0 & 0 & 0 \\ 0 & 0 & 1 & 0 \\ 0 & 1 & 0 & 0 \\ 0 & 0 & 0 & 1 \end{pmatrix} = (s_N, s_2, s_1, s_3)$$

$$\Gamma(\sigma_v'') \quad (s_N, s_1, s_2, s_3)\begin{pmatrix} 1 & 0 & 0 & 0 \\ 0 & 0 & 0 & 1 \\ 0 & 0 & 1 & 0 \\ 0 & 1 & 0 & 0 \end{pmatrix} = (s_N, s_3, s_2, s_1)$$

These six matrices therefore form a representation for the C_{3v} point group in the (s_N, s_1, s_2, s_3) basis. They multiply together according to the group multiplication table and satisfy all the requirements for a mathematical group.

Example: a matrix representation of the C_{2v} point group (the allyl radical) In this example, we'll take as our basis a p orbital on each carbon atom (p_1, p_2, p_3).

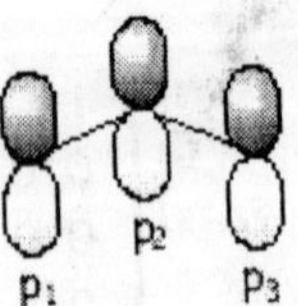

The symmetry operations in the C_{2v} point group, and their effect on the three p orbitals, are as follows:

$$E\ (p_1, p_2, p_3) \rightarrow (p_1, p_2, p_3)$$

$$C_2\ (p_1, p_2, p_3) \rightarrow (-p_3, -p_2, -p_1)$$

$$\sigma_v\ (p_1, p_2, p_3) \rightarrow (-p_1, -p_2, -p_3)$$

$$\sigma_v'(p_1, p_2, p_3) \rightarrow (p_3, p_2, p_1)$$

The matrices that carry out the transformation are

$$\Gamma(E) \qquad (p_1,p_2,p_3)\begin{pmatrix} 1 & 0 & 0 \\ 0 & 1 & 0 \\ 0 & 0 & 1 \end{pmatrix} = (p_1,p_2,p_3)$$

$$\Gamma(C_2) \qquad (p_1,p_2,p_3)\begin{pmatrix} 0 & 0 & -1 \\ 0 & -1 & 0 \\ -1 & 0 & 0 \end{pmatrix} = (-p_3,-p_2,-p_1)$$

$$\Gamma(\sigma_v) \qquad (p_1,p_2,p_3)\begin{pmatrix} -1 & 0 & 0 \\ 0 & -1 & 0 \\ 0 & 0 & -1 \end{pmatrix} = (-p_1,-p_2,-p_3)$$

$$\Gamma(\sigma_v') \qquad (p_1,p_2,p_3)\begin{pmatrix} 0 & 0 & 1 \\ 0 & 1 & 0 \\ 1 & 0 & 0 \end{pmatrix} = (p_3,p_1,p_2)$$

Properties of Matrix Representations

Now that we have learnt how to create a matrix representation of a point group within a given basis, we will move on to look at some of the properties that make these representations so powerful in the treatment of molecular symmetry.

Similarity Transforms

Suppose we have a basis set $(x_1, x_2, x_3, \ldots x_n)$, and we have determined the matrix representatives for the basis in a given point group. There is nothing particularly special about the basis set we have chosen, and we could equally have used any set of linear combinations of the original functions (provided the combinations were linearly independent).

The matrix representatives for the two basis sets will be certainly different, but we would expect them to be related to each other in some way. As we shall show shortly, they are in fact related by a similarity transform. It will be far from obvious at this point why we would want to carry out such a transformation, but similarity transforms will become important later on when we use group theory to choose an optimal basis set with which to generate molecular orbitals.

Consider a basis set $(x_1', x_2', x_3', \ldots x_n')$, in which each basis function x_i' is a linear combination of our original basis $(x_1, x_2, x_3, \ldots x_n)$.

$$x_j' = \Sigma_i\, x_i c_{ji} = x_1 c_{j1} + x_2 c_{j2} + \ldots$$

The c_{ji} appearing in the sum are coefficients; c_{ji} is the coefficient multiplying the original basis function x_i in the new linear combination basis function x_j'. We could also represent this transformation in terms of a matrix equation $x' = xC$:

$$(x_1',x_2',...x_n') = (x_1,x_2,...x_n)\begin{pmatrix} c_{11} & c_{12} & \cdots & c_{1n} \\ c_{21} & c_{22} & \cdots & c_{2n} \\ \cdots & \cdots & \cdots & \cdots \\ c_{n1} & c_{n2} & \cdots & c_{nn} \end{pmatrix}$$

Now we look at what happens when we apply a symmetry operation g to our two basis sets. If $\Gamma(g)$ and $\Gamma(g)$ are matrix representatives of the symmetry operation in the x and x' bases, then we have:

$gx' = x'\,\Gamma(g)$

$gxC = xC\Gamma(g)$ since $x' = xC$

$gx = xC\ \Gamma(g)C^{-1}$ multiplying on the right by C^{-1} and using $CC^{-1} = I$

$= x\ G\ (g)$

We can therefore identify the similarity transform relating $\Gamma(g)$, the matrix representative in our original basis, to $\Gamma'(g)$, the representative in the transformed basis. The transform depends only on the matrix of coefficients used to transform the basis functions.

$$\Gamma(g) = C\ \Gamma'(g)C^{-1}$$

$$\text{Also } \Gamma'(g) = C^{-1}\ \Gamma(g)C$$

Characters of Representations

The trace of a matrix representative $\Gamma(g)$ is usually referred to as the character of the representation under the symmetry operation g. We will soon come to see that the characters of a matrix representation are often more useful than the matrix representatives themselves. Characters have several important properties.

1. The character of a symmetry operation is invariant under a similarity transform.

2. Symmetry operations belonging to the same class have the same character in a given representation. Note that the character for a given class may be different in different representations, and that more than one class may have the same character.

Reduction of Representations I

Let us now go back and look at the C_{3v} representation. If we look at the matrices carefully we see that they all take the same block diagonal form (a square matrix is said to be block diagonal if all the elements are zero except for a set of submatrices lying along the diagonal).

$$\Gamma(E) = \begin{pmatrix} 1 & 0 & 0 & 0 \\ 0 & 1 & 0 & 0 \\ 0 & 0 & 1 & 0 \\ 0 & 0 & 0 & 1 \end{pmatrix} \quad \Gamma(C_3^+) = \begin{pmatrix} 1 & 0 & 0 & 0 \\ 0 & 0 & 0 & 1 \\ 0 & 1 & 0 & 0 \\ 0 & 0 & 1 & 0 \end{pmatrix} \quad \Gamma(C_3^-) = \begin{pmatrix} 1 & 0 & 0 & 0 \\ 0 & 0 & 1 & 0 \\ 0 & 0 & 0 & 1 \\ 0 & 1 & 0 & 0 \end{pmatrix}$$

$$\chi(E) = 4 \qquad \chi(C_3^+) = 1 \qquad \chi(C_3^-) = 1$$

$$\Gamma(\sigma_v) = \begin{pmatrix} 1 & 0 & 0 & 0 \\ 0 & 1 & 0 & 0 \\ 0 & 0 & 0 & 1 \\ 0 & 0 & 1 & 0 \end{pmatrix} \quad \Gamma(\sigma_v') = \begin{pmatrix} 1 & 0 & 0 & 0 \\ 0 & 0 & 1 & 0 \\ 0 & 1 & 0 & 0 \\ 0 & 0 & 0 & 1 \end{pmatrix} \quad \Gamma(\sigma_v'') = \begin{pmatrix} 1 & 0 & 0 & 0 \\ 0 & 0 & 0 & 1 \\ 0 & 0 & 1 & 0 \\ 0 & 1 & 0 & 0 \end{pmatrix}$$

$$\chi(\sigma_v) = 2 \qquad \chi(\sigma_v') = 2 \qquad \chi(\sigma_v'') = 2$$

A block diagonal matrix can be written as the direct sum of the matrices that lie along the diagonal. In the case of the C_{3v} matrix representation, each of the matrix representatives may be written as the direct sum of a 1×1 matrix and a 3×3 matrix.

$$\Gamma^{(4)}(g) = \Gamma^{(1)}(g) \oplus \Gamma^{(3)}(g)$$

The bracketed superscripts denote the dimensionality of the matrices. Note that a direct sum is very different from ordinary matrix addition since it produces a matrix of higher dimensionality. A direct sum of two matrices of orders n and m is performed by placing the matrices to be summed along the diagonal of a matrix of order $n + m$ and filling in the remaining elements with zeroes.

The reason why this result is useful in group theory is that the two sets of matrices $\Gamma^{(1)}(g)$ and $\Gamma^{(3)}(g)$ also satisfy all of the requirements for a matrix representation. Each set contains the identity and an inverse for each member, and the members multiply together associatively according to the group multiplication table. Recall that the basis for the original four-dimensional representation had the s orbitals (s_N, s_1, s_2, s_3) of ammonia as its basis. The first set of reduced matrices, $\Gamma(1)(g)$, forms a one-dimensional representation with (s_N) as its basis. The second set, $\Gamma^{(3)}(g)$ forms a three-dimensional representation with the basis (s_1, s_2, s_3). Separation of the original representation into representations of lower dimensionality is called reduction of the representation. The two reduced representations are shown below.

g	E	C_3^+	C_3^-	σ_v	σ_v'	σ_v''	
$\Gamma^{(1)}(g)$	(1)	(1)	(1)	(1)	(1)	(1)	*1D representation sponned by (S_N)*
$\Gamma^{(3)}(g)$	$\begin{pmatrix}1&0&0\\0&1&0\\0&0&1\end{pmatrix}$	$\begin{pmatrix}0&0&1\\1&0&0\\0&1&0\end{pmatrix}$	$\begin{pmatrix}0&1&0\\0&0&1\\1&0&0\end{pmatrix}$	$\begin{pmatrix}1&0&0\\0&0&1\\0&1&0\end{pmatrix}$	$\begin{pmatrix}0&1&0\\1&0&0\\0&0&1\end{pmatrix}$	$\begin{pmatrix}0&0&1\\0&1&0\\1&0&0\end{pmatrix}$	*3D representation sponned by (s_1, s_2, s_3)*

The logical next step is to investigate whether or not the three dimensional representation $\Gamma^{(3)}(g)$ can be reduced any further. As it stands, the matrices making up this representation are not in block diagonal form so the representation is not reducible. However, we can carry out a similarity transformation to a new representation spanned by a new set of basis functions (made up of linear combinations of (s_1, s_2, s_3)), which is reducible. In this case, the appropriate (normalised) linear combinations to use as our new basis functions are

$$s_1' = \frac{1}{\sqrt{3}}(s_1 + s_2 + s_3)$$

$$s_2' = \frac{1}{\sqrt{6}}(2s_1 - s_2 - s_3)$$

$$s_3' = \frac{1}{\sqrt{2}}(s_2 - s_3)$$

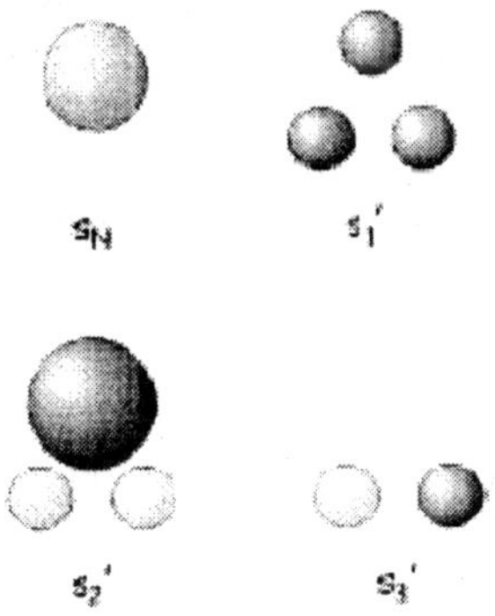

or in matrix form

$$\underset{X'}{(s_1', s_2', s_3')} = \underset{X}{(s_1, s_2, s_3)} \underset{C}{\begin{pmatrix} 1/\sqrt{3} & 2/\sqrt{6} & 0 \\ 1/\sqrt{3} & -1/\sqrt{6} & 1/\sqrt{2} \\ 1/\sqrt{3} & -1/\sqrt{6} & -1/\sqrt{2} \end{pmatrix}}$$

The matrices in the new representation are found from $\Gamma'(g) = C^{-1}\Gamma(g)C$ to be

$$\Gamma^{(3)\prime}(g) \quad \underset{E}{\begin{pmatrix} 1 & 0 & 0 \\ 0 & 1 & 0 \\ 0 & 0 & 1 \end{pmatrix}} \quad \underset{C_3^+}{\begin{pmatrix} 1 & 0 & 0 \\ 0 & -1/2 & -\sqrt{3}/2 \\ 0 & \sqrt{3}/2 & -1/2 \end{pmatrix}} \quad \underset{C_3^-}{\begin{pmatrix} 1 & 0 & 0 \\ 0 & -1/2 & \sqrt{3}/2 \\ 0 & -\sqrt{3}/2 & -1/2 \end{pmatrix}}$$

$$\underset{\sigma_v}{\begin{pmatrix} 1 & 0 & 0 \\ 0 & 1 & 0 \\ 0 & 0 & -1 \end{pmatrix}} \quad \underset{\sigma_v'}{\begin{pmatrix} 1 & 0 & 0 \\ 0 & -1/2 & \sqrt{3}/2 \\ 0 & \sqrt{3}/2 & 1/2 \end{pmatrix}} \quad \underset{\sigma_v''}{\begin{pmatrix} 1 & 0 & 0 \\ 0 & -1/2 & -\sqrt{3}/2 \\ 0 & -\sqrt{3}/2 & 1/2 \end{pmatrix}}$$

We see that each matrix is now in block diagonal form, and the representation may be reduced into the direct sum of a 1×1 representation spanned by (s_1') and a 2×2 representation spanned by (s_2', s_3'). The complete set of reduced representations obtained from the original 4D representation is:

E	C_3^+	C_3^-	σ_v	σ_v'	σ_v''	
(1)	(1)	(1)	(1)	(1)	(1)	1D representation spanned by (s_N)
(1)	(1)	(1)	(1)	(1)	(1)	1D representation spanned by (s_1')

$$\begin{pmatrix} 1 & 0 \\ 0 & 1 \end{pmatrix} \begin{pmatrix} -1/2 & -\sqrt{3}/2 \\ \sqrt{3}/2 & -1/2 \end{pmatrix} \begin{pmatrix} -1/2 & \sqrt{3}/2 \\ -\sqrt{3}/2 & -1/2 \end{pmatrix}$$

$$\begin{pmatrix} 1 & 0 \\ 0 & -1 \end{pmatrix} \begin{pmatrix} -1/2 & \sqrt{3}/2 \\ \sqrt{3}/2 & 1/2 \end{pmatrix} \begin{pmatrix} -1/2 & -\sqrt{3}/2 \\ -\sqrt{3}/2 & 1/2 \end{pmatrix}$$

2D representation spanned by (s_1', s_3')

This is as far as we can go in reducing this representation. None of the three representations above can be reduced any further, and they are therefore called irreducible representations, or 'irreps', of the point group. Formally, a representation is an irrep if there is no similarity transform that can simultaneously convert all of the representatives into block diagonal form. The linear combination of basis functions that converts a matrix representation into block diagonal form, allowing reduction of the representation, is called a symmetry adapted linear combination.

Irreducible Representations and Symmetry Species

The two one-dimensional irreps spanned by s_N and s_1' are seen to be identical. This means that s_N and s_1' have the 'same symmetry', transforming in the same way under all of the symmetry operations of the point group and forming bases for the same matrix representation. As such, they are said to belong to the same symmetry species. There are a limited number of ways in which an arbitrary function can transform under the symmetry operations of a group, giving rise to a limited number of symmetry species. Any function that forms a basis for a matrix representation of a group must transform as one of the symmetry species of the group. The irreps of a point group are labelled according to their symmetry species as follows:

i) 1D representations are labelled A or B, depending on whether they are symmetric (character +1) or antisymmetric (character –1) under rotation about the principal axis.

ii) 2D representations are labelled E, 3D representations are labelled T.

iii) In groups containing a centre of inversion, g and u labels (from the German gerade and ungerade, meaning symmetric and antisymmetric) denote the character of the irrep under inversion (+1 for g, –1 for u)

iv) In groups with a horizontal mirror plane but no centre of inversion, the irreps are given prime and double prime labels to denote whether they are symmetric (character +1) or antisymmetric (character –1) under reflection in the plane.

v) If further distinction between irreps is required, subscripts 1 and 2 are used to denote the character with respect to a C_2 rotation perpendicular to the principal axis, or with respect to a vertical reflection if there are no C_2 rotations.

The 1D irrep in the C_{3v} point group is symmetric (has character +1) under all the symmetry operations of the group. It therefore belongs to the irrep A_1. The 2D irrep has character 2 under the identity operation, –1 under rotation, and 0 under reflection, and belongs to the irrep E.

Character Tables

Character tables summarise the behaviour of all of the possible irreps of a group under each of the symmetry operations of the group. The character table for C_{3v} is shown below.

C_{3v}, $3m$	E	$2C_3$	$3\sigma_v$	$h=6$
A_1	1	1	1	z, z^2, x^2+y^2
A_2	1	1	-1	R_z
E	2	-1	0	(x, y), (xy, x^2+y^2), (xz, yz), (R_x, R_y)

The various sections of the table are as follows:

i) The first element in the table gives the name of the point group, usually in both Schoenflies (C_{3v}) and Hermann-Mauguin (3*m*) notation.

ii) Along the first row are the symmetry operations of the group, E, $2C_3$ and $3\sigma_v$, followed by the order h of the group. Because operations in the same class have the same character, symmetry operations are grouped into classes in the character table and not listed separately.

iii) In the first column are the irreps of the group. In C_{3v} the irreps are A_1, A_2 and E (the representation we considered above spans $2A_1 + E$).

iv) The characters of the irreps under each symmetry operation are given in the bulk of the table.

v) The final column of the table lists a number of functions that transform as the various irreps of the group. These are the Cartesian axes (x, y, z) the Cartesian products (z^2, $x^2 + y^2$, xy, xz, yz) and the rotations (R_x, R_y, R_z).

Though we would not be using them much in this course, the functions listed in the final column of the table are important in many chemical applications of group theory, particularly in spectroscopy. For example, by looking at the transformation properties of *x*, *y* and *z* (sometimes given in character tables as T_x, T_y, T_z) we can discover the symmetry of translations along the x, y, and z axes. Similarly, R_x, R_y and R_z represent rotations about the three Cartesian axes. The transformation properties of x, y, and z can also be used to determine whether or not a molecule can absorb a photon of *x*–, *y*– or *z*– polarised light and undergo a spectroscopic transition. The Cartesian products play as similar role in determining selection rules for Raman transitions, which involve two photons.

Reduction of representations II

As we shall see later, the formation of chemical bonds is strongly dependent on the atomic orbitals involved having the correct symmetries. Group theory is extremely useful in the treatment of chemical bonding, since by making maximum use of molecular symmetry, we greatly simplify the process of constructing molecular orbitals and determining their energies. In order to take maximum advantage of symmetry in this way, we need to develop a little more 'machinery'. Specifically, given a basis set of atomic orbitals, we need to find out:

1. How to determine the irreps spanned by the basis functions.
2. How to construct linear combinations of the original basis functions that transform as a given irrep/symmetry species.

It turns out that both of these problems can be solved using something called the 'Great Orthogonality Theorem' (GOT for short). The GOT summarises a number of orthogonality relationships implicit in matrix representations of symmetry groups, and may be derived in a somewhat qualitative fashion by considering these relationships in turn.

Orthogonality Relationships in Group Theory

1. If corresponding matrix elements in all of the matrix

representatives of an irrep are squared and added together, the result is equal to the order of the group divided by the dimensionality of the irrep. i.e.

$$\Sigma_g \Gamma_k(g)_{ij} \Gamma_k(g)_{ij} = h/dk \quad (1)$$

where k labels the irrep, i and j label the row and column position within the irrep, h is the order of the group, and d_k is the order of the irrep, e.g. The order of the group C_{3v} is 6. If we apply the above operation to the first element in the 2 × 2 (E) the result should be equal to $h/d_k = 6/2 = 3$. Carrying out this operation gives:

$$(1)^2 + (-½)^2 + (-½)^2 + (1)^2 + (-½)^2 + (-½)^2$$
$$= 1 + ¼ + ¼ + 1 + ¼ + ¼ = 3$$

2. If instead of summing the squares of matrix elements in an irrep, we sum the product of two different elements from within each matrix, the result is equal to zero, i.e.

$$\Sigma g \, \Gamma_k(g)_{ij} \Gamma_k(g)_{i'j}' = 0 \quad (2)$$

where i ≠ i' and/or $j \neq j'$. e.g. if we perform this operation using the two elements in the first row of the 2D irrep used in 1, we get:

$$(1)(0) + (-\tfrac{1}{2})(-\tfrac{\sqrt{3}}{2}) + (-\tfrac{1}{2})(\tfrac{\sqrt{3}}{2}) + (1)(0) + (-\tfrac{1}{2})(\tfrac{\sqrt{3}}{2}) + (-\tfrac{1}{2})(-\tfrac{\sqrt{3}}{2})$$

$$= 0 + \frac{\sqrt{3}}{4} - \frac{\sqrt{3}}{4} + 0 - \frac{\sqrt{3}}{4} + \frac{\sqrt{3}}{4} = 0$$

3. If we sum the product of two elements from the matrices of two different irreps k and m, the result is equal to zero. i.e.

$$\Sigma_g \Gamma_k(g)_{ij} \Gamma_m(g)_{i'j'} = 0 \quad (3)$$

where there is now no restriction on the values of the indices i, j, i', j' (apart from the rather obvious restriction that they must be less than or equal to the dimensions of the irrep). e.g. Performing this operation on the first elements of the A_1 and E irreps we derived for C_{3v} gives:

$$(1)(1) + (1)(-\tfrac{1}{2}) + (1)(-\tfrac{1}{2}) + (1)(1) + (1)(-\tfrac{1}{2}) + (1)(-\tfrac{1}{2}) = 1 - \tfrac{1}{2} - \tfrac{1}{2} + 1 - \tfrac{1}{2} - \tfrac{1}{2} = 0$$

We can combine these three results into one general equation, the Great Orthogonality Theorem.

$$\Sigma_g \Gamma_k(g)_{ij} \Gamma_m(g)_{i'j'} = \frac{h}{\sqrt{d_k d_m}} \delta_{km}\delta_{ii'}\delta_{jj'} \quad (4)$$

For most applications we do not actually need the full Great Orthogonality Theorem. A little mathematical trickery transforms Equation (4) into the 'Little Orthogonality Theorem' (or LOT), which is expressed in terms of the characters of the irreps rather than the irreps themselves.

$$\Sigma_g \chi_k(g) \chi_m(g) = h\delta_{km} \quad (5)$$

Since the characters for two symmetry operations in the same class are the same, we can also rewrite the sum over symmetry operations as a sum over classes.

$$\Sigma_C n_C \chi_k(C) \chi_m(C) = h\delta_{km} \quad (6)$$

where n_C is the number of symmetry operations in class C. In all of the examples we have considered so far, the characters have been real. However, this is not necessarily true for all point groups, so to make the above equations completely general we need to include the possibility of imaginary characters. In this case we have:

$$\Sigma_c n_c \chi_k^*(C)\chi_m(C) = h\delta_{km} \qquad (7)$$

where $\chi_k^*(C)$ is the complex conjugate of $\chi_k(C)$. Equation (7) is of course identical to (6) when all the characters are real.

Using the LOT to Determine the Irreps Spanned by B basis

We discovered that we can often carry out a similarity transform on a general matrix representation so that all the representatives end up in the same block diagonal form. When this is possible, each set of submatrices also forms a valid matrix representation of the group. If none of the submatrices can be reduced further by carrying out another similarity transform, they are said to form an irreducible representation of the point group. An important property of matrix representatives is that their character is invariant under a similarity transform. This means that the character of the original representatives must be equal to the sum of the characters of the irreps into which the representation is reduced, e.g, if we consider the representative for the C_3^+ symmetry operation in our NH_3 example, we have:

$$\begin{pmatrix} 1 & 0 & 0 & 0 \\ 0 & 0 & 0 & 1 \\ 0 & 1 & 0 & 0 \\ 0 & 0 & 1 & 0 \end{pmatrix} \xrightarrow{\text{similarity transform}} \begin{pmatrix} 1 & 0 & 0 & 0 \\ 0 & 1 & 0 & 0 \\ 0 & 0 & -1/2 & -\sqrt{3}/2 \\ 0 & 0 & \sqrt{3}/2 & -1/2 \end{pmatrix}$$

$$\chi = 1 \qquad\qquad \chi = 1$$

$$= (1) \oplus (1) \oplus \begin{pmatrix} -1/2 & -\sqrt{3}/2 \\ \sqrt{3}/2 & -1/2 \end{pmatrix}$$

$$\chi = 1 + 1 + -1 = 1$$

It follows that we can write the characters for a general representation $\Gamma(g)$ in terms of the characters of the irreps $\Gamma_k(g)$ into which it can be reduced.

$$\chi(g) = \Sigma_k\ a_k\ \chi_k(g) \tag{8}$$

where the coefficients a_k in the sum are the number of times each irrep appears in the representation. This means that in order to determine the irreps spanned by a given basis. all we have to do is determine the coefficients a_k in the above equation. This is where the Little Orthogonality Theorem comes in handy. If we take the LOT in the form of Equation 5, and multiply each side through by a_k, we get

$$\Sigma_g\ a_k \chi_k(g)\ \chi_m(g) = h\ a_k \delta_{km} \tag{9}$$

Summing both sides of the above equation over k gives

$$\Sigma_g\ \Sigma_k\ a_k \chi_k(g)\ \chi_m(g) = h\ \Sigma_k\ a_k \delta^k{}_m$$

We can use Equation (8) to simplify the left hand side of this equation. Also, the sum on the right hand side reduces to am because δ_{km} is only non-zero (and equal to 1) when $k = m$

$$\Sigma_g\ \chi(g)\ \chi_m(g) = h\ a_m$$

Dividing both sides through by h (the order of the group), gives us an expression for the coefficients a_m in terms of the characters $\chi(g)$ of the original representation and the characters $\chi_m(g)$ of the mth irrep.

$$a_m\ 1/h\ \Sigma_g\ \chi(g)\ \chi_m(g) \tag{10}$$

We can of course write this as a sum over classes rather than a sum over symmetry operations.

$$a_m = 1/h\ \Sigma_c n_c(g) \chi_m(g) \tag{11}$$

As an example, the matrix representatives we derived for the C_{3v} group could be reduced into two irreps of A_1 symmetry and one of E symmetry. i.e. $\Gamma = 2A_1 + E$. We could have obtained the same result using Equation (10). The characters for our original representation and for the irreps of the C_{3v} point group (A_1, A_2 and E) are given in the table below.

C_{3v}	E	$2C_3$	$3\sigma_v$
χ	4	1	2
$\chi(A_1)$	1	1	1
$\chi(A_2)$	1	1	-1
$\chi(E)$	2	-1	0

From (11), the number of times each irrep occurs for our chosen basis (s_N, s_1, s_2, s_3) is therefore

$$a(A_1) = \frac{1}{6}\,(\,1\text{x}4\text{x}1 + 2\text{x}1\text{x}1 + 3\text{x}2\text{x}1\,) = 2$$

$$a(A_2) = \frac{1}{6}\,(\,1\text{x}4\text{x}1 + 2\text{x}1\text{x}1 + 3\text{x}2\text{x}{-1}\,) = 0$$

$$a(E) = \frac{1}{6}\,(\,1\text{x}4\text{x}2 + 2\text{x}1\text{x}{-1} + 3\text{x}2\text{x}0\,) = 1$$

i.e. basis is spanned by $2A_1 + E$, as we found before.

Symmetry Adapted Linear Combinations (SALC)

Once we know the irreps spanned by an arbitrary basis set, we can work out the appropriate linear combinations of basis functions that transform the matrix representatives of our original representation into block diagonal form (i.e. the symmetry adapted linear combinations). Each of the SALCs transforms as one of the irreps of the reduced representation. We have already seen this in our NH_3 example. The two linear combinations of A_1 symmetry were s_N and $s_1 + s_2 + s_3$, both of which are symmetric under all the symmetry operations of the point group. We also chose another pair of functions, $2s_1 - s_2 - s_3$ and $s_2 - s_3$, which together transform as the symmetry species E.

To find the appropriate SALCs to reduce a matrix representation, we use projection operators. You will be familiar with the idea of operators from quantum mechanics. The operators we will be using here are not quantum mechanical operators, but the basic principle is the same. The projection operator to generate a SALC that transforms as an irrep k is $\Sigma_g \chi_k(g)g$. Each term in the sum means 'apply the symmetry operation g and then multiply by the character of g in irrep k'. Applying this operator to each of our original basis functions in turn will generate a complete set of SALCs. i.e. to transform a basis function f_i into a SALC f_i', we use

$$f_i' = \Sigma_g \chi_k(g)\ g\ f_i \qquad (12)$$

The way in which this operation is carried out will become much more clear if we work through an example. We can break down the above equation into a fairly straightforward 'recipe' for generating SALCs:

1. Make a table with columns labelled by the basis functions and rows labelled by the symmetry operations of the molecular point group. In the columns, show the effect of the symmetry operations on the basis functions (this is the $g\ f_i$ part of Equation (12)).
2. For each irrep in turn, multiply each member of the table by the character of the appropriate symmetry operation (we now have $\chi_k(g)\ g\ f_i$ for each operation). Summing over the columns (symmetry operations) generates all the SALCs that transform as the chosen irrep.
3. Normalise the SALCs.

Earlier, we worked out the effect of all the symmetry operations in the C_{3v} point group on the s_N, s_1, s_2, s_3) basis.

E	$(s_N, s_1, s_2, s_3) \rightarrow (s_N, s_1, s_2, s_3)$
C_3^+	$(s_N, s_1, s_2, s_3) \rightarrow (s_N, s_3, s_1, s_2)$

$$C_3^- (s_N, s_1, s_2, s_3) \rightarrow (s_N, s_2, s_3, s_1)$$
$$\sigma_v (s_N, s_1, s_2, s_3) \rightarrow (s_N, s_1, s_3, s_2)$$
$$\sigma_v' (s_N, s_1, s_2, s_3) \rightarrow (s_N, s_2, s_1, s_3)$$
$$\sigma_v'' (s_N, s_1, s_2, s_3) \rightarrow (s_N, s_3, s_2, s_1)$$

This is all we need to construct the table described in 1. above.

	s_N	s_1	s_2	s_3
E	s_N	s_1	s_2	s_3
C_3^+	s_N	s_3	s_1	s_2
C_3^-	s_N	s_2	s_3	s_1
σ_v	s_N	s_1	s_3	s_2
σ_v'	s_N	s_2	s_1	s_3
σ_v''	s_N	s_3	s_2	s_1

To determine the SALCs of A_1 symmetry, we multiply the table through by the characters of the A_1 irrep (all of which take the value 1). Summing the columns gives

$$s_N + s_N + s_N + s_N + s_N + s_N = 6s_N$$
$$s_1 + s_3 + s_2 + s_1 + s_2 + s_3 = 2(s_1 + s_2 + s_3)$$
$$s_2 + s_1 + s_3 + s_3 + s_1 + s_2 = 2(s_1 + s_2 + s_3)$$
$$s_3 + s_2 + s_1 + s_2 + s_3 + s_1 = 2(s_1 + s_2 + s_3)$$

Apart from a constant factor (which doesn't affect the functional form and therefore doesn't affect the symmetry properties), these are the same as the combinations we determined earlier. Normalising gives us two SALCs of A_1 symmetry.

$$\phi_1 = s_N$$
$$\phi_2 = 1\sqrt{3}\ (s_1 + s_2 + s_3)$$

We now move on to determine the SALCs of E symmetry. Multiplying the table above by the appropriate characters for the E irrep gives

	s_N	s_1	s_2	s_3
E	$2s_N$	$2s_1$	$2s_2$	$2s_3$
C_3^+	$-s_N$	$-s_3$	$-s_1$	$-s_2$
C_3^-	s_N	$-s_2$	$-s_3$	$-s_1$
σ_v	0	0	0	0
σ_v'	0	0	0	0
σ_v''	0	0	0	0

Summing the columns yields

$$2s_N - s_N - s_N = 0$$
$$2s_1 - s_3 - s_2$$
$$2s_2 - s_1 - s_3$$
$$2s_3 - s_2 - s_1$$

We therefore get three SALCs from this procedure. This is a problem, since the number of SALCs must match the dimensionality of the irrep, in this case two. Put another way, we should end up with four SALCs in total to match our original number of basis functions. Added to our two SALCs of A_1 symmetry, three SALCs of E symmetry would give us five in total.

The resolution to our problem lies in the fact that the three SALCs above are not linearly independent. Any one of them can be written as a linear combination of the other two e.g. $(2s_1-s_3-s_2) = -(2s_2-s_1-s_3) - (2s_3-s_2-s_1)$. To solve the problem, we can either throw away one of the SALCs, or better, make two linear combinations of the three SALCs that are orthogonal to each other. e.g. if we take $2s_1 - s_2 - s_3$ as one of our SALCs and find an orthogonal combination of the other two (which turns out to be their difference), we have (after normalisation)

$$\phi_3 = \frac{1}{\sqrt{6}}(2s_1 - s_2 - s_3)$$

$$\phi_4 = \frac{1}{\sqrt{2}}(s_2 - s_3)$$

We are now ready to apply group theory to the problem of constructing molecular orbitals.

Bonding in Diatomics

You will already be familiar with the idea of constructing molecular orbitals from linear combinations of atomic orbitals from previous courses covering bonding in diatomic molecules. By considering the symmetries of *s* and *p* orbitals on two atoms, we can form bonding and antibonding combinations labelled as having either *s* or *p* symmetry, depending on whether they resemble *s* or *p* orbitals when viewed along the bond axis. In all of the cases shown, only atomic orbitals that have the same symmetry when viewed along the bond axis z can form a chemical bond e.g. two *s* orbitals, two p_z orbitals, or an *s* and a p_z can form a bond, but a p_z and a p_x or an *s* and a p_x or a p_y cannot. It turns out that the rule that determines whether or not two atomic orbitals can bond is that they must belong to the same symmetry species within the point group of the molecule. We can prove this mathematically for two atomic orbitals ϕ_i and ϕ_j by looking at the overlap integral between the two orbitals.

$$S_{ij} = \langle\phi_i|\phi_j\rangle = \int\phi_i^*\phi_j\, dt$$

In order for bonding to be possible, this integral must be non-zero. The product of the two functions ϕ_1 and ϕ_2 transforms as the direct product of their symmetry species, i.e. $\Gamma_{12} = \Gamma_1 \otimes \Gamma_2$. Because an integral is just a number, it must be invariant to any symmetry operation and must therefore transform as the totally symmetric irrep in the appropriate point group. This means that $\Gamma_1 \otimes \Gamma_2$ must include the totally symmetric irrep. As it happens, this is only possible if ϕ_1 and ϕ_2 belong to the same irrep. These ideas are summarised for a diatomic in the table below.

First atomic orbital		*Second atomic orbital*		$\Gamma_1 \otimes \Gamma_2$	*Overlap integral*	*Bonding?*
s	(A_{1g})	s	(A_{1g})	A_{1g}	Non-zero	Yes
s	(A_{1g})	p_x	(E_{1u})	E_{1u}	Zero	No
s	(A_{1g})	p_z	(A_{1u})	A_{1u}	Zero	No
p_x	(E_{1u})	p_x	(E_{1u})	A_{1g} + A_{2g} + E_{2g}	Non-zero	Yes
p_x	(E_{1u})	p_z	(A_{1u})	E_{1g}	Zero	No
p_z	(A_{1u})	p_z	(A_{1u})	A_{1g}	Non-zero	Yes

Bonding in polyatomics - constructing molecular orbitals from SALCs

How to use symmetry to determine whether two atomic orbitals can form a chemical bond. How do we carry out the same procedure for a polyatomic molecule, in which many atomic orbitals may combine to form a bond? Any SALCs of the same symmetry could potentially form a bond, so all we need to do to construct a molecular orbital is take a linear combination of all the SALCs of the same symmetry species. The general procedure is:

1. Use a basis set consisting of valence atomic orbitals on each atom in the system.
2. Determine which irreps are spanned by the basis set and construct the SALCs that transform as each irrep.
3. Take linear combinations of irreps of the same symmetry species to form the molecular orbitals.

e.g. in NH_3 example we could form a molecular orbital of A_1 symmetry from the two SALCs that transform as A_1,

$$\Psi(A_1) = c_1\ \phi_1 + c_2\ \phi_2$$
$$= c_1 s_N + c_2\ 1/\sqrt{3}\ (s_1 + s_2 + s_3) \qquad (13)$$

Unfortunately, this is as far as group theory can take us. It can give us the functional form of the molecular orbitals but it cannot determine the coefficients c_1 and c_2. To go further and obtain the expansion coefficients and orbital energies, we must turn to quantum mechanics.

The Variation Principle

Molecular energy levels, or orbital energies, are eigenvalues of the molecular Hamiltonian $\hat{H}$ The energy E of a molecular orbital Y may therefore be written

$$E = \frac{\langle\Psi|\hat{H}|\Psi\rangle}{\langle\Psi|\Psi\rangle} \quad \text{(unnormalised } \Psi\text{)} \tag{14}$$

or $E = \langle\Psi|\hat{H}|\Psi\rangle$ (normalised Ψ, for which $\langle\Psi|\Psi\rangle=1$) (15)

The variation principle states that any approximate wavefunction must have a higher energy than the true wavefunction. This follows directly from the fairly common-sense idea that in general any system tries to minimize its energy. If an 'approximate' wavefunction had a lower energy than the 'true' wavefunction, we would expect the system to try and adopt this 'approximate' lower energy state, rather than the 'true' state. That all approximations to the true wavefunction must have a higher energy than the true wavefunction is the only scenario that makes physical sense. The idea behind the variation principle is very straightforward, but when constructing molecular orbitals from SALCs it turns out to be all we need to determine both the unknown expansion coefficients and the orbital energies. If the true wavefunction has the lowest energy, then to find the closest approximation to the true wavefunction, all we have to do is to find the coefficients in expansion of SALCs that minimise the energy. In practice, we substitute wavefunction into above Equation and minimise the resulting expression with respect to the coefficients. Substituting into Equation (14) above gives:

$$E = \frac{\langle c_1\phi_1 + c_2\phi_2 | \hat{H} | c_1\phi_1 + c_2\phi_2 \rangle}{\langle c_1\phi_1 + c_2\phi_2 | c_1\phi_1 + c_2\phi_2 \rangle}$$

$$= \frac{\langle c_1\phi_1 | \hat{H} | c_1\phi_1 \rangle + \langle c_1\phi_1 | \hat{H} | c_2\phi_2 \rangle + \langle c_2\phi_2 | \hat{H} | c_1\phi_1 \rangle + \langle c_2\phi_2 | \hat{H} | c_2\phi_2 \rangle}{\langle c_1\phi_1 | c_1\phi_1 \rangle + \langle c_1\phi_1 | c_2\phi_2 \rangle + \langle c_2\phi_2 | c_1\phi_1 \rangle + \langle c_2\phi_2 | c_2\phi_2 \rangle}$$

$$= \frac{c_1^2\langle \phi_1 | \hat{H} | \phi_1 \rangle + c_1c_2\langle \phi_1 | \hat{H} | \phi_2 \rangle + c_2c_1\langle \phi_2 | \hat{H} | \phi_1 \rangle + c_2^2\langle \phi_2 | \hat{H} | \phi_2 \rangle}{c_1^2\langle \phi_1 | \phi_1 \rangle + c_1c_2\langle \phi_1 | \phi_2 \rangle + c_2c_1\langle \phi_2 | \phi_1 \rangle + c_2^2\langle \phi_2 | \phi_2 \rangle}$$

If we now define a Hamiltonian matrix element $H_{ij} = \langle \phi_i | \hat{H} | \phi_j \rangle$ and an overlap integral $S_{ij} = \langle \phi_i | \phi_j \rangle$ and note that $H_{ij} = H_{ji}$ and $S_{ij} = S_{ji}$, this simplifies to

$$E = \frac{c_1^2H_{11} + 2c_1c_2H_{12} + c_2^2H_{22}}{c_1^2S_{11} + 2c_1c_2S_{12} + c_2^2S_{22}}$$

To get this into a simpler form, we multiply both sides through by the denominator to give

$$E\,(c_1^2S_{11} + 2c_1c_2S_{12} + c_2^2S_{22}) = c_1^2H_{11} + 2c_1c_2H_{12} + c_2^2H_{22}$$

Now we need to minimise the energy with respect to c_1 and c_2 i.e. we require $\delta E/\delta c_1 = 0$ and $\delta E/\delta c_2 = 0$. If we differentiate the above equation through separately by c_1 and c_2 and apply this condition, we will end up with two equations in the two unknowns c_1 and c_2, which we can solve to determine the coefficients and the energy.

Differentiating by c_1 gives

$$\frac{\partial E}{\partial c_1}(c_1^2S_{11} + 2c_1c_2S_{12} + c_2^2S_{22}) + E(2c_1S_{11} + 2c_2S_{12}) = 2c_1H_{11} + 2c_2H_{12}$$

Differentiating by c_2 gives

$$\frac{\partial E}{\partial c_2}(c_1^2S_{11} + 2c_1c_2S_{12} + c_2^2S_{22}) + E(2c_1S_{12} + 2c_2S_{22}) = 2c_1H_{12} + 2c_2H_{22}$$

Because $\dfrac{\partial E}{\partial c_1} = \dfrac{\partial E}{\partial c_2} = 0,$

The first term on the left hand side of both equations is zero, leaving us with

$$E(2c_1S_{11} + 2c_2S_{12}) = 2c_1H_{11} + 2c_2H_{12}$$
$$E(2c_1S_{12} + 2c_2S_{22}) = 2c_1H_{12} + 2c_2H_{22}$$

These are normally rewritten slightly, in the form

$$c_1(H_{11} - ES_{11}) + c_2(H_{12} - ES_{12}) = 0$$
$$c_1(H_{12} - ES_{12}) + c_2(H_{22} - ES_{22}) = 0 \qquad (16)$$

These equations are known as the secular equations and are the set of equations we need to solve to determine c_1, c_2 and E. In the general case, when our wavefunction is a linear combination of N SALCs we get N equations in N unknowns, with the kth equation given by

$$\Sigma_{i=1}^{N} \; c_i(H_{ik} - ES_{ik}) = 0 \qquad (17)$$

Note that we can use any basis functions we like together with the linear variation method described here to construct approximate molecular orbitals and determine their energies, but choosing to use SALCs simplifies things considerably. As we shall see in the following section, an arbitrary set of N basis functions leads to a set of N equations in N unknowns, which must be solved simultaneously. Converting our basis into a set of SALCs separates the equations into several smaller sets of secular equations, one for each irrep, which can be solved independently. It is usually easier to solve several sets of secular equations of lower dimensionality than one set of higher dimensionality.

Solving the Secular Equations

Matrix Formulation of a Set of Linear Equations

As we have seen already, any set of linear equations may be rewritten as a matrix equation $Ax = b$. Linear equations are classified as simultaneous linear equations or homogeneous linear equations, depending on whether the vector b on the RHS of the equation is non-zero or zero. For a set of simultaneous linear equations (non-zero b) it is fairly apparent that if a unique solution exists, it can be found by multiplying both sides by the inverse matrix A^{-1} (since $A^{-1}A$ on the left hand side is equal to the identity matrix, which has no effect on the vector x)

$$Ax = b$$

$$\mathrm{A}^{-1}Ax = A^{-1}b$$

$$x = A^{-1}b$$

In practice, there are easier matrix methods for solving simultaneous equations than finding the inverse matrix, but these need not concern us here. We discovered that in order for a matrix to have an inverse, it must have a non-zero determinant. Since A^{-1} must exist in order for a set of simultaneous linear equations to have a solution, this means that the determinant of the matrix A must be non-zero for the equations to be solvable. The reverse is true for homogeneous linear equations. In this case the set of equations only has a solution if the determinant of A is equal to zero. The secular equations we want to solve are homogeneous equations, and we will use this property of the determinant to determine the molecular orbital energies. An important property of homogeneous equations is that if a vector x is a solution, so is any multiple of x, meaning that the solutions can be normalised without causing any problems.

Solving for the orbital energies and expansion coefficients

Recall the secular equations for the A_1 orbitals of NH_3.

$$c_1(H_{11}-ES_{11}) + c_2(H_{12}-ES_{12}) = 0$$
$$c_1(H_{12}-ES_{12}) + c_2(H_{22}-ES_{22}) = 0$$

where c_1 and c_2 are the coefficients in the linear combination of the SALCs $\phi_1 = s_N$ and $\phi_2 = 1/\sqrt{3}\ (s_1 + s_2 + s_3)$ used construct the molecular orbital. Writing this set of homogeneous linear equations in matrix form gives

$$\begin{matrix} H_{11}-ES_{11} & H_{12}-ES_{12} \\ H_{12}-ES_{12} & H_{22}-ES_{22} \end{matrix} \begin{pmatrix} c_1 \\ c_2 \end{pmatrix} = \begin{pmatrix} 0 \\ 0 \end{pmatrix} \tag{18}$$

In order for the equations to have a solution, the determinant of the matrix must be equal to zero. Writing out the determinant will give us a polynomial equation in E that we can solve to obtain the orbital energies in terms the Hamiltonian matrix elements H_{ij} and overlap integrals S_{ij}. The number of energies obtained by solving the secular determinant in this way is equal to the order of the matrix, in this case two.

The secular determinant for Equation above is (noting that $S_{11} = S_{22} = 1$ since the SALCs are normalised)

$$(H_{11}-E)(H_{22}-E) - (H_{12}-ES_{12})^2 = 0$$

Expanding and collecting terms in E gives

$$E^2(1-S_{12}^{\ 2}) + E(2H_{12}S_{12}-H_{11}-H_{22}) + (H_{11}H_{22}-H_{12}^{\ 2}) = 0$$

$E\pm$ =

$$\frac{(2H_{12}S_{12} - H_{11} - H_{22} \pm \sqrt{\begin{matrix} 2H_{12}S_{12} - H_{11} - H_{22})^2 - \\ 4(1 - S_{12}^{\ 2})H_{11}H_{22} - H_{22}^{\ 2}) \end{matrix}}}{2(1 - S_{12}^{\ 2})} \tag{19}$$

To obtain numerical values for the energies, we need to evaluate the integrals H_{11}, H_{22}, H_{12}, S_{12}. This would be quite a challenge to do analytically, but luckily there are a number of computer programs that can be used to calculate the integrals. One such program gives the following values.

$$H_{11} = -26.0000 \text{ eV}$$
$$H_{22} = -22.2216 \text{ eV}$$
$$H_{12} = -29.7670 \text{ eV}$$
$$S_{12} = 0.8167$$

When we substitute these into our equation for the energy levels, we get:

$$E+ = 29.8336 \text{ eV}$$
$$E- = -31.0063 \text{ eV}$$

We now have the orbital energies. The next step is to find the orbital coefficients. The coefficients for an orbital of energy E are found by substituting the energy into the secular equations and solving for the coefficients c_i. Since the two secular equations are not linearly independent (i.e. they are effectively only one equation), when we solve them to find the coefficients what we will end up with is the relative values of the coefficients. This is true in general in a system with N coefficients, solving the secular equations will allow all N of the coefficients c_i to be obtained in terms of, say, c_1. The absolute values of the coefficients are found by normalising the wavefunction.

Since the secular equations for the orbitals of energy E_+ and E_- are not linearly independent, we can choose to solve either one of them to find the orbital coefficients. We will choose the first.

$$(H_{11} - E_{\pm})c_1 + (H_{12} - E_{\pm}S_{12})c_2 = 0$$

For the orbital with energy $E_- = -31.0063$ eV, substituting numerical values into this equation gives

$$5.0063\ c_1 - 4.4442\ c_2 = 0$$
$$c_2 = 1.1265\ c_1$$

The molecular orbital is therefore

$$\psi_1 = c_1(\phi_1 + 1.1265\ \phi_2)$$

Normalising to find the constant c_1 (by requiring $\langle\psi|\psi\rangle = 1$) gives

$$\psi_1 = 0.4933\ \phi_1 + 0.5557\ \phi_2$$
$$= 0.4933\ s_N + 0.3208\ (s_1 + s_2 + s_3)$$

(substituting the SALCs for ϕ_1 and ϕ_2)

For the second orbital, with energy $E_+ = 29.8336$ eV, the secular equation is

$$-55.8336\ c_1 - 54.1321\ c_2 = 0$$
$$c_2 = -1.0314\ c_1$$

giving

$$\psi_2 = c_1(\phi_1 - 1.0314\ \phi_2)$$
$$= 1.6242\ \phi_1 - 1.6752\ \phi_2 \text{ (after normalisation)}$$
$$= 1.6242\ s_N - 0.9672\ (s_1 + s_2 + s_3)$$

These two A_1 molecular orbitals ψ_1 and ψ_2, one bonding and one antibonding, are shown below.

The remaining two SALCs arising from the s orbitals of NH_3 ($\phi_3 = \frac{1}{\sqrt{6}}(2s_1 - s_2 - s_3)$ and $\phi_4 = \frac{1}{\sqrt{2}}(s_2 - s_3)$), form an orthogonal pair of molecular orbitals of E symmetry. We can show this by solving the secular determinant to find the orbital energies. The secular equations in this case are:

$$\begin{pmatrix} H_{33}-ES_{33} & H_{34}-ES_{34} \\ H_{34}-ES_{34} & H_{44}-ES_{44} \end{pmatrix} \begin{pmatrix} c_3 \\ c_3 \end{pmatrix} = \begin{pmatrix} 0 \\ 0 \end{pmatrix}$$

Solving the secular determinant gives

$$E\pm = \frac{-(2H_{34} - S_{34} - H_{33} - H_{44} \pm \sqrt{\begin{array}{l}(2H_{34}S_{34} - H_{33} - H_{44})^2 - \\ 4(1-S_{34})^2(H_{33}H_{44} - H_{34}{}^2)\end{array}}}{2(1-S_{34}{}^2)}$$

The integrals required are

$$H_{33} = -9.2892 \text{ eV}$$
$$H_{44} = -9.2892 \text{ eV}$$
$$H_{34} = 0$$
$$S_{34} = 0$$

Using the fact that $H_{34} = S_{34} = 0$, the expression for the energies reduces to

$$E\pm = \frac{(H_{33} + H_{44}) + (H_{33} + H_{44})}{2}$$

giving $E_+ = H_{33} = -9.2892$ eV and $E_- = H_{44} = -9.2892$ eV. Each SALC therefore forms a molecular orbital by itself, and the two orbitals have the same energy; the two SALCs form an orthogonal pair of degenerate orbitals.

These two molecular orbitals of E symmetry are shown below.

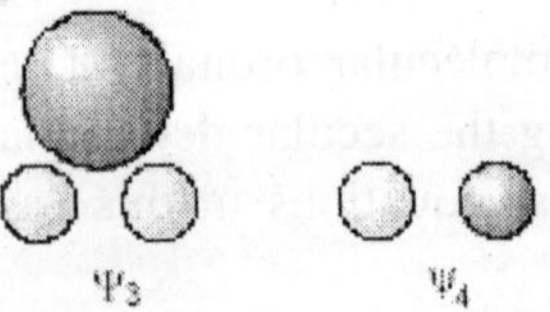

Huckel Molecular Orbital Theory

Calculation of the integrals H_{ii}, H_{ij} and S_{ij} needed to determine the orbital coefficients and energies is not generally straightforward and is usually done numerically on a computer. However, in some cases we can come up with approximate treatments that allow analytical solutions to be obtained. Huckel theory is a simplified version of the linear variation method that can be applied to conjugated π systems. The first approximation is to assume $\sigma - \pi$ separability, which means that the nuclei, inner shell electrons and localised σ bonds provide an effective field in which the remaining π electrons move. The effective field is assumed to be the same for each carbon atom.

The next simplification is to approximate the p molecular orbitals as linear combinations of atomic orbitals. In a minimal basis set calculation of a planar conjugated hydrocarbon, the only atomic orbitals with π symmetry are the $2p$ orbitals on carbon. The π orbitals can therefore be written

$$\Psi_i = \sum_{i=1}^{N} c_{ij} \, \mathrm{f}_j$$

where f_j is a $2p\pi$ orbital on the jth carbon atom. The optimum values of the coefficients c_{ij} and the orbital energies for the N lowest lying molecular orbitals can then be found by solving the secular equations. The key assumptions in Huckel theory involve the integrals H_{ij} and S_{ij}.

i) Assuming that H_{ii} has the same value (α) for every carbon atom in the molecule, and also for carbon atoms in different planar hydrocarbons. This is fairly reasonable since the electronic structure of all the conjugated atoms is similar.

ii) The integral H_{ij} is assumed to have the same value (β) for any two carbon atoms bonded to each other and to vanish for two non-bonded atoms. Since non-bonded carbon atoms are usually well separated and therefore unlikely to interact very much, this is also a reasonable assumption.

ii) The overlap integral S_{ij} is assumed to be equal to zero unless $i = j$, in which case it is equal to unity since the atomic orbitals f_i are normalised. Assuming an overlap integral of zero for bonded atoms is obviously a poor approximation (the whole basis of a chemical bond is atomic orbital overlap!) and in practice the theory is often modified to improve on this assumption. Even so, the approximation tends to shift the orbital energies in a fairly straightforward way and doesn't change the relative ordering of energy levels too much. The Huckel approximation may be summarised as follows:

$$H_{ii} = \langle f_i|\mathrm{H}^{\mathrm{eff}}|f_i\rangle \equiv a$$
$$H_{ij} = \langle f_i|\mathrm{H}^{\mathrm{eff}}|f_j\rangle \equiv \beta \text{ for atoms } i, j \text{ bonded}$$
$$H_{ij} = \langle f_i|H^{\mathrm{eff}}|f_j\rangle = 0 \text{ for atoms } i, j \text{ not bonded}$$
$$S_{ij} = \langle f_i|f_j\rangle = \delta_{ij}$$

α and β are called the Coulomb and bond (sometimes resonance) integrals. β is the important characteristic parameter of Huckel theory, since it defines the spacing of the molecular energy levels. Note that both α and β are negative.

The molecular orbital coefficients can be used to determine a range of molecular properties. As an example, one quantity that is sometimes quoted in relation to Huckel theory is the π electron or charge density on an atom. The charge density on atom j is given by $q_j = S_i\, n_i|c_{ij}|^2$. While this is not a true electron density, measuring neither charge nor probability density, but merely the number of π electrons in the vicinity of atom i, it can sometimes be used to give at least a rough indication of the reactivity of particular atoms with nucleophiles and electrophiles.

Example—*the molecular orbitals of the allyl radical*

We constructed a matrix representation for the allyl radical using the basis (p_1, p_2, p_3) consisting of the three p orbitals, one on each atom, that make up the conjugated π system. We can use this to construct SALCs and molecular orbitals. The matrix representatives and their characters are:

$$E \quad \begin{pmatrix} 1 & 0 & 0 \\ 0 & 1 & 0 \\ 0 & 0 & 1 \end{pmatrix} \qquad C_2 \quad \begin{pmatrix} 0 & 0 & -1 \\ 0 & -1 & 0 \\ -1 & 0 & 0 \end{pmatrix} \qquad \sigma_v \quad \begin{pmatrix} -1 & 0 & 0 \\ 0 & -1 & 0 \\ 0 & 0 & -1 \end{pmatrix} \qquad \sigma_v' \quad \begin{pmatrix} 0 & 0 & 1 \\ 0 & 1 & 0 \\ 1 & 0 & 0 \end{pmatrix}$$

$$\chi(E) = 3 \qquad \chi(C_2) = -1 \qquad \chi(\sigma_v) = -3 \qquad \chi(\sigma_v') = 1$$

The first thing to do is to determine which irreps of the C_{2v} point group are spanned by the basis, using

$a_k = \frac{1}{h}\, \Sigma_C\, n_C\, \chi(g)\, \chi_k(g)$ to find the number of times each irrep appears. The character table for the C_{2v} point group is

C_{2v}	E	C_2	σ_v	σ_v'	$h = 4$
A_1	1	1	1	1	z, x^2, y^2, z^2
A_2	1	1	−1	−1	xy, R_z
B_1	1	−1	1	−1	x, xz, R_y
B_2	1	−1	−1	1	y, yz, R_x

We have

$$a(A_1) = \tfrac{1}{4}\ (1 \times 3 \times 1 + 1 \times -1 \times 1 + 1 \times -3 \times 1 + 1 \times 1 \times 1) = 0$$

$$a(A_2) = \tfrac{1}{4}\ (1 \times 3 \times 1 + 1 \times -1 \times 1 + 1 \times -3 \times -1 + 1 \times 1 \times -1) = 1$$

$$a(B_1) = \tfrac{1}{4}\ (1 \times 3 \times 1 + 1 \times -1 \times -1 + 1 \times -3 \times 1 + 1 \times 1 \times -1) = 0$$

$$a(B_2) = \tfrac{1}{4}\ (1 \times 3 \times 1 + 1 \times -1 \times -1 + 1 \times -3 \times -1 + 1 \times 1 \times 1) = 2$$

so the basis spans $A_2 + 2B_2$. Now we form our SALCs by applying the projection operator

$$\phi_i = \Sigma_g\ \chi_k(g)\ g\ f_i$$

to each of the basis functions in turn. The effect of acting on the (p_1, p_2, p_3) basis with the symmetry operations in the C_{2v} group is

$$E\ (p_1, p_2, p_3) \rightarrow (p_1, p_2, p_3)$$
$$C_2\ (p_1, p_2, p_3) \rightarrow (-p_3, -p_2, -p_1)$$
$$\sigma_v\ (p_1, p_2, p_3) \rightarrow (-p_1, -p_2, -p_3)$$
$$\sigma_v'\ (p_1, p_2, p_3) \rightarrow (p_3, p_2, p_1)$$

First we will find the SALCs of B_2 symmetry. Applying the projection operator to the p_1, p_2 and p_3 basis functions we get

$$\phi(p_1) = (1 \times p_1) + (-1 \times -p_3) + (-1 \times -p_1) + (1 \times p_3) = 2(p_1 + p_3)$$

$$\phi(p_2) = (1 \times p_2) + (-1 \times -p_2) + (-1 \times -p_2) + (1 \times p_2) = 4p_2$$

$$\phi(p_3) = (1 \times p_3) + (-1 \times -p_1) + (-1 \times -p_3) + (1 \times p_1) = 2(p_1 + p_3) = \phi(p_1)$$

After normalisation, the two SALCs of B_2 symmetry are

$$\phi_1 = \frac{1}{\sqrt{2}}\,(p_1 - p_3)$$

$$\phi_2 = p_2$$

Now for the SALC of A_2 symmetry. From the three p basis functions we get

$$\phi(p_1) = (1 \times p_1) + (1 \times -p_3) + (-1 \times -p_1) + (-1 \times p_3) = 2(p_1 - p_3)$$

$$\phi(p_2) = (1 \times p_2) + (1 \times -p_2) + (-1 \times -p_2) + (-1 \times p_2) = 0$$

$$\phi(p_3) = (1 \times p_3) + (1 \times -p_1) + (-1 \times -p_3) + (-1 \times p_1) = 2(p_3 - p_1) = -\phi(p_1)$$

$\phi(p_1)$ and $\phi(p_3)$ are not linearly independent so we can take either as our third SALC. We choose

$$\phi_3 = \frac{1}{\sqrt{2}} (p_1 - p_3)$$

Since we have three atomic orbitals, we can construct three molecular orbitals, two of B_2 symmetry and one of A_2 symmetry, by taking linear combinations of SALCs of the same symmetry species. The SALC of A_2 symmetry does not combine with any others, so one of the molecular orbitals is simply

$$\psi(A_2) = \phi_3 = \frac{1}{\sqrt{2}} (p_1 - p_3)$$

We can find the energy $E(A_2)$ of this orbital within the Huckel approximation. There is only one secular equation

$$H_{33} - E(B_1)S_{33} = 0$$

$$E(B_1) = H_{33}/S_{33}$$

The matrix elements are:

$$S_{33} = \langle\phi_3|\phi_3\rangle = 1$$

$$H_{33} = \langle\phi_3|H|\phi_3\rangle$$

$$= \tfrac{1}{2} \langle p_1 - p_3|H|p_1 - p_3\rangle$$

$$= \tfrac{1}{2} (\langle p_1|H|p_1\rangle - \langle p_1|H|p_3\rangle - \langle p_3|H|p_1\rangle + \langle p_3|\mathrm{H}|p_3\rangle)$$

$$= \tfrac{1}{2} (\alpha - 0 - 0 + a)$$

$$= \alpha$$

The orbital energy is therefore $E(A_2) = \alpha$. The two orbitals of B_2 symmetry are given by linear combinations of ϕ_1 and ϕ_2 (with different coefficients for the two orbitals):

$$\psi_1(B_2) = c_1\phi_1 + c_2\phi_2$$
$$\psi_2(B_2) = c_1'\phi_1 + c_2'\phi_2$$

In this case, the secular equations are:

$$(H_{11}-ES_{11})c_1 + (H_{21} - ES_{21})c_2 = 0$$
$$(H_{12}-ES_{12})c_1 + (H_{22} - ES_{22})c_2 = 0$$

The matrix elements are:

$$S_{11} = S_{22} = 1$$
$$S_{12} = S_{21} = <\phi_1|\phi_2>$$
$$= \frac{1}{\sqrt{2}} <p_1+p_3|p_2>$$
$$= \frac{1}{\sqrt{2}} (<p_1|p_2> + <p_3|p_2>)$$
$$= \alpha$$

$$H_{11} = <f_1|H|\frac{1}{\sqrt{2}}\ \frac{1}{\sqrt{2}}\ {}_1>$$
$$= \tfrac{1}{2} <p_1 + p_3|H|p_1 + p_3>$$
$$= \tfrac{1}{2} (<p_1|H|p_1> + <p_1|H|p_3> + <p_3|H|p_1> + <p_3|H|p_3>)$$
$$= \tfrac{1}{2} (\alpha + 0 + 0 + a)$$
$$= \alpha$$

$$\mathrm{H}_{22} = <f_2|H|f_2>$$
$$= <p_2|H|p_2>$$
$$= \alpha$$

$$H_{12} = H_{21} = <\phi_1|H|\phi_2>$$

$$= \frac{1}{\sqrt{2}} <p_1+p_3|H|p_2>$$

$$= \frac{1}{\sqrt{2}} (<p_1|H|p_2> + <p_3|H|p_2>)$$

$$= \frac{1}{\sqrt{2}} (\beta + \beta)$$

$$= \sqrt{2}\ \beta$$

Substituting into the secular equations gives

$$(\alpha\text{-}E)c_1 + \sqrt{2}\beta\ c_2 = 0$$
$$\sqrt{2}\beta\ c_1 + (\alpha - E)c_2 = 0$$

Solving the secular determinant to find the orbital energies gives:

$$\det \begin{pmatrix} \alpha - E & \sqrt{2}\beta \\ \sqrt{2}\beta & \alpha - \mathrm{E} \end{pmatrix} = (\alpha - E)^2 - 2\beta^2 = 0$$

$$(\alpha - E)^2 = 2\beta^2$$

$$\alpha - \mathrm{E} = \pm \sqrt{2}\beta$$

$$E = \alpha \pm 2\beta$$

We find the coefficients for the SALCs making up the molecular orbitals by substituting these energies back into the secular equations and solving. Either of the secular equations will give the same result. Using the first equation gives:

$$[\alpha - (\alpha \pm \sqrt{2}\beta)]c_1 + \sqrt{2}\beta\ c_2 = 0$$

$$c_2 = \pm\ c_1$$

and the two orbitals are

$$\psi_1(B_2) = c_1(\phi_1 + \phi_2)$$

$$= \frac{1}{\sqrt{2}}(\phi_1 + \phi_2)$$

$$= \tfrac{1}{2}(p_1 + \sqrt{2}\, p_2 + p_3)$$

$$\psi_2(B_2) = c_1(\phi_1 - \phi_2)$$

$$= \frac{1}{\sqrt{2}}(\phi_1 - \phi_2)$$

$$= \tfrac{1}{2}(p_1 - \sqrt{2}\, p_2 + p_3)$$

In summary, our three Huckel molecular orbitals for the allyl radical are:

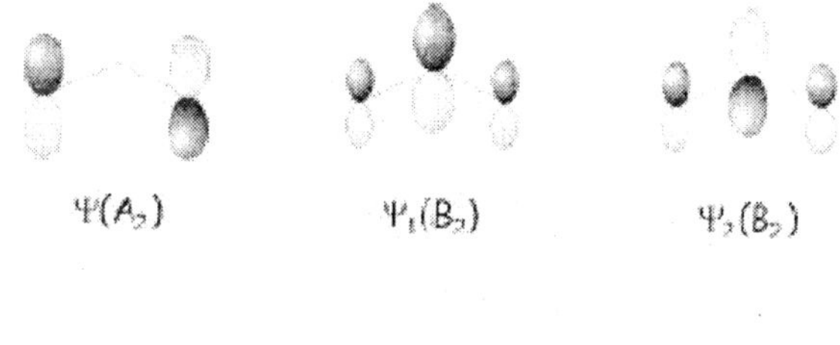

$$\psi(A_2) = \frac{1}{\sqrt{2}}(p_1 - p_3)$$

$$\psi_1(B_2) = \tfrac{1}{2}(p_1 + \sqrt{2}\, p_2 + p_3)$$

$$\psi_2(B_2) = \tfrac{1}{2}(p_1 - \sqrt{2}\, p_2 + p_3)$$

Example: The Molecular Orbitals of H_2O

As another example, we will use group theory to construct the molecular orbitals of H_2O (point group C_{2v}) using a basis set consisting of all the valence orbitals. The valence orbitals are a 1s orbital on each hydrogen, which we will label s_H and s_H', and a 2s and three 2p orbitals on the oxygen, which we will label s_O, p_x, p_y, p_z, giving a complete basis (s_H, s_H', s_O, p_x, p_y, p_z).

The first thing to do is to determine how each orbital transforms under the symmetry operations of the C_{2v} point group (E, C_2, σ_v and σ_v'), construct a matrix representation and determine the characters of each operation. The symmetry operations and axis system we will be using are shown below.

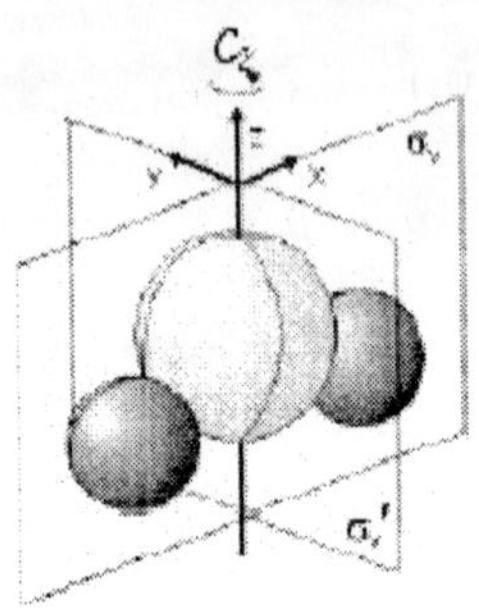

The orbitals transform in the following way

$$E\ (s_H, s_H', s_O, p_x, p_y, p_z) \rightarrow (s_H, s_H', s_O, p_x, p_y, p_z)$$

$$C_2\ (s_H, s_H', s_O, p_x, p_y, p_z) \rightarrow (s_H', s_H, s_O, -p_x, -p_y, p_z)$$

$$\sigma_v(_{xz})\ (s_H, s_H', s_O, p_x, p_y, p_z) \rightarrow (s_H, s_H', s_O, p_x, -p_y, p_z)$$

$$\sigma_v'(_{yz})\ (s_H, s_H', s_O, p_x, p_y, p_z) \rightarrow (s_H', s_H, s_O, -p_x, p_y, p_z)$$

A short aside on constructing matrix representatives

After a little practice, you will probably be able to write matrix representatives straight away just by looking at the effect of the symmetry operations on the basis. However, if you are struggling a little the following procedure might help.

Remember that the matrix representatives are just the matrices we would have to multiply the left hand side of the above equations by to give the right hand side. In most cases they are very easy to work out. Probably the most straightforward way to think about it is that each column of the matrix shows where one of the original basis functions ends up. For example, the first column transforms the basis function s_H to its new position. The first column of the matrix can be found by taking the result on the right hand side of the above expressions, replacing every function that isn't s_H with a zero, putting the coefficient of s_H (1 or –1 in this example) in the position at which it occurs, and taking the transpose to give a column vector. e.g. Consider the representative for the C_2 operation.

The original basis (s_H, s_H', s_O, p_x, p_y, p_z) transforms into (s_H', s_H, s_O, $-p_x$, $-p_y$, p_z). The first column of the matrix therefore transforms s_H into s_H'. Taking the result and replacing all the other functions with zeroes gives (0, s_H, 0, 0, 0, 0). The coefficient of s_H is 1, so the first column of the C_2 matrix representative is

$$\begin{pmatrix} 0 \\ 1 \\ 0 \\ 0 \\ 0 \\ 0 \end{pmatrix}$$

Matrix Representation, Characters and SALCs

The matrix representatives and their characters are

$$E \qquad\qquad\qquad\qquad C_2$$

$$\begin{pmatrix} 1 & 0 & 0 & 0 & 0 & 0 \\ 0 & 1 & 0 & 0 & 0 & 0 \\ 0 & 0 & 1 & 0 & 0 & 0 \\ 0 & 0 & 0 & 1 & 0 & 0 \\ 0 & 0 & 0 & 0 & 1 & 0 \\ 0 & 0 & 0 & 0 & 0 & 1 \end{pmatrix} \qquad \begin{pmatrix} 0 & 1 & 0 & 0 & 0 & 0 \\ 1 & 0 & 0 & 0 & 0 & 0 \\ 0 & 0 & 1 & 0 & 0 & 0 \\ 0 & 0 & 0 & -1 & 0 & 0 \\ 0 & 0 & 0 & 0 & -1 & 0 \\ 0 & 0 & 0 & 0 & 0 & 1 \end{pmatrix}$$

$$\chi(E) = 6 \qquad\qquad\qquad\qquad \chi(C_2) = 0$$

$$\sigma_v \quad \begin{pmatrix} 1 & 0 & 0 & 0 & 0 & 0 \\ 0 & 1 & 0 & 0 & 0 & 0 \\ 0 & 0 & 1 & 0 & 0 & 0 \\ 0 & 0 & 0 & 1 & 0 & 0 \\ 0 & 0 & 0 & 0 & -1 & 0 \\ 0 & 0 & 0 & 0 & 0 & 1 \end{pmatrix} \qquad \sigma_v' \quad \begin{pmatrix} 0 & 1 & 0 & 0 & 0 & 0 \\ 1 & 0 & 0 & 0 & 0 & 0 \\ 0 & 0 & 1 & 0 & 0 & 0 \\ 0 & 0 & 0 & -1 & 0 & 0 \\ 0 & 0 & 0 & 0 & 1 & 0 \\ 0 & 0 & 0 & 0 & 0 & 1 \end{pmatrix}$$

$$\chi(\sigma_v) = 4 \qquad \chi(\sigma_v') = 2$$

Now we are ready to work out which irreps are spanned by the basis we have chosen. The character table for C_{2v} is:

C_{2v}	E	C_2	σ_v	σ_v'	$h = 4$
A_1	1	1	1	1	z, x^2, y^2, z^2
A_2	1	1	-1	-1	xy, R_z
B_1	1	-1	1	-1	x, xz, R_y
B_2	1	-1	-1	1	y, yz, R_x

As before, we use Equation to find out the number of times each irrep appears.

$$a_k = \frac{1}{h} \Sigma_C \, n_C \chi(g) \, \chi_k(g)$$

We have

$$a(A_1) = ¼ \, (1 \times 6 \times 1 + 1 \times 0 \times 1 + 1 \times 4 \times 1 + 1 \times 2 \times 1) = 3$$

$$a(A_2) = ¼ \, (1 \times 6 \times 1 + 1 \times 0 \times 1 + 1 \times 4 \times -1 + 1 \times 2 \times -1) = 0$$

$$a(B_1) = ¼ \, (1 \times 6 \times 1 + 1 \times 0 \times -1 + 1 \times 4 \times 1 + 1 \times 2 \times -1) = 2$$

$$a(B_2) = ¼ \, (1 \times 6 \times 1 + 1 \times 0 \times -1 + 1 \times 4 \times -1 + 1 \times 2 \times 1) = 1$$

so the basis spans $3A_1 + 2B_1 + B_2$. Now we use the projection operators applied to each basis function ϕ_i in turn to determine the SALCs $\phi_i = \Sigma_g \, \chi_k(g) \, g \, f_i$

The SALCs of A_1 symmetry are:

$$\phi(s_H) = s_H + s_H' + s_H + s_H' = 2(s_H + s_H')$$

$$\phi(s_H') = s_H' + s_H + s_H' + s_H = 2(s_H + s_H')$$

$$\phi(s_O) = s_O + s_O + s_O + s_O = 4s_O$$

$$\phi(p_x) = p_x - p_x + p_x - p_x = 0$$

$$\phi(p_y) = p_y - p_y - p_y + p_y = 0$$

$$\phi(p_z) = p_z + p_z + p_z + p_z = 4p_z$$

The SALCs of B_1 symmetry are:

$$\phi(s_H) = s_H - s_H' + s_H - s_H' = 2(s_H - s_H')$$

$$\phi(s_H') = s_H' - s_H + s_H' - s_H = 2(s_H' - s_H)$$

$$\phi(s_O) = s_O - s_O + s_O - s_O = 0$$

$$\phi(p_x) = p_x + p_x + p_x + p_x = 4p_x$$

$$\phi(p_y) = p_y + p_y - p_y - p_y = 0$$

$$\phi(p_z) = p_z - p_z + p_z - p_z = 0$$

The SALCs of B_2 symmetry are:

$$\phi(s_H) = s_H - s_H' - s_H + s_H' = 0$$

$$\phi(s_H') = s_H' - s_H - s_H' + s_H = 0$$

$$\phi(s_O) = s_O - s_O - s_O + s_O = 0$$

$$\phi(p_x) = p_x + p_x - p_x - p_x = 0$$

$$\phi(p_y) = p_y + p_y + p_y + p_y = 4p_y$$

$$\phi(p_z) = p_z - p_z - p_z + p_z = 0$$

After normalisation, our SALCs are therefore:

A_1 symmetry

$$\phi_1 = \frac{1}{\sqrt{2}}(s_H + s_H')$$

$\phi_2 = s_O$

$\phi_3 = p_z$

B_1 symmetry

$$\phi_4 = \frac{1}{\sqrt{2}}(s_H - s_H')$$

$\phi_5 = p_x$

B_2 symmetry

$\phi_6 = p_y$

Note that we only take one of the first two SALCs generated by the B_1 projection operator since one is a simple multiple of the other (i.e. they are not linearly independent). We can therefore construct three molecular orbitals of A_1 symmetry, with the general form

$$\begin{aligned}\psi(A_1) &= c_1\,\phi_1 + c_2\,\phi_2 + c_3\,\phi_3 \\ &= c_1'(s_H + s_H') + c_2 s_O + c_3 p_z \text{ where } c_1' = c_1/\sqrt{2}\end{aligned}$$

two molecular orbitals of B_1 symmetry, of the form

$$\begin{aligned}\psi(B_1) &= c_4\,\phi_4 + c_5\,\phi_5 \\ &= c_4'(s_H - s_H') + c_5 p_z\end{aligned}$$

and one molecular orbital of B_2 symmetry

$$\begin{aligned}\psi(B_2) &= \phi_6 \\ &= p_y\end{aligned}$$

To work out the coefficients $c_1 - c_5$ and determine the orbital energies, we would have to solve the secular equations for each set of orbitals in turn. We are not dealing with a conjugated π

system, so in this case Huckel theory cannot be used and the various H_{ij} and S_{ij} integrals would have to be calculated numerically and substituted into the secular equations. This involves a lot of tedi ous algebra, which we will leave out for the moment. The LCAO orbitals determined above are an approximation of the true molecular orbitals of water, which are shown on the right. As we have shown using group theory, the A_1 molecular orbitals involve the oxygen $2s$ and $2p_z$ atomic orbitals and the sum $s_H + s_H'$ of the hydrogen 1s orbitals. The B_1 molecular orbitals involve the oxygen $2p_x$ orbital and the difference $s_H - s_H'$ of the two hydrogen 1s orbitals, and the B_2 molecular orbital is essentially an oxygen $2p_y$ atomic orbital.

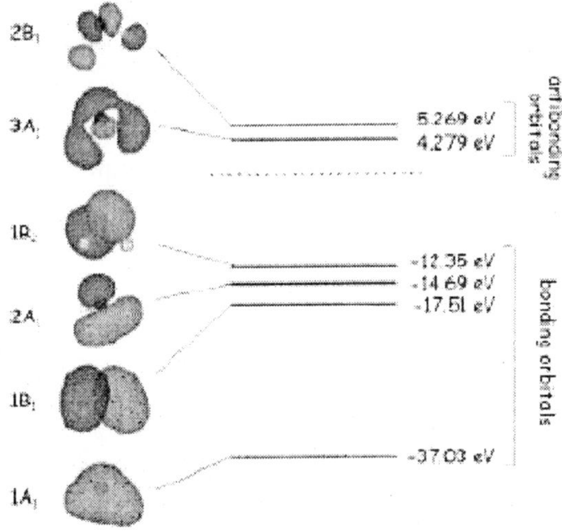

Example : The Molecular Orbitals of s-Trans Butadiene

As a final example, we will use group theory to construct the π orbitals for *s*-trans butadiene, and determine their energy ordering using Huckel molecular orbital theory. *s*-trans butadiene belongs to the point group C_{2h}. Our basis set consists of a p orbital on each carbon atom (p_1, p_2, p_3, p_4) perpendicular to the molecular plane, as shown below.

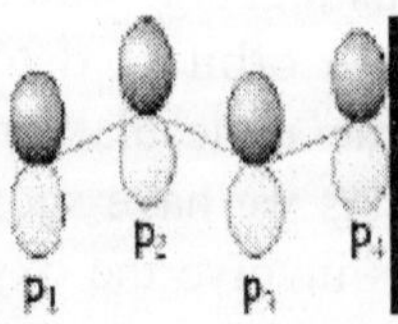

C_{2h}	E	C_2	i	σ_h	$h = 4$
A_g	1	1	1	1	
B_g	1	-1	1	-1	
A_u	1	1	-1	-1	
B_u	1	-1	-1	1	

Hopefully by now you're becoming familiar with the procedure for constructing molecular orbitals. Again, the first thing to do is to work out how each of our basis functions transform under the four symmetry operations and construct the matrix representation.

$$E\,(p_1, p_2, p_3, p_4) \rightarrow (p_1, p_2, p_3, p_4)$$
$$C_2\,(p_1, p_2, p_3, p_4) \rightarrow (p_4, p_3, p_2, p_1)$$
$$i\,(p_1, p_2, p_3, p_4) \rightarrow (-p_4, -p_3, -p_2, -p_1)$$
$$\sigma_h(p_1, p_2, p_3, p_4) \rightarrow (-p_1, -p_2, -p_3, -p_4)$$

The matrix representatives are therefore

$$\begin{array}{cccc} E & C_z & i & \sigma_n \\ \begin{pmatrix} 1 & 0 & 0 & 0 \\ 0 & 1 & 0 & 0 \\ 0 & 0 & 1 & 0 \\ 0 & 0 & 0 & 1 \end{pmatrix} & \begin{pmatrix} 0 & 0 & 0 & 1 \\ 0 & 0 & 1 & 0 \\ 0 & 1 & 0 & 0 \\ 1 & 0 & 0 & 0 \end{pmatrix} & \begin{pmatrix} 0 & 0 & 0 & -1 \\ 0 & 0 & -1 & 0 \\ 0 & -1 & 0 & 0 \\ -1 & 0 & 0 & 0 \end{pmatrix} & \begin{pmatrix} -1 & 0 & 0 & 0 \\ 0 & -1 & 0 & 0 \\ 0 & 0 & -1 & 0 \\ 0 & 0 & 0 & -1 \end{pmatrix} \\ \chi(E) = 4 & \chi(C_2) = 4 & \chi(i) = 0 & \chi(\sigma n) = -4 \end{array}$$

Now we work out which irreps are spanned by the (p_1, p_2, p_3, p_4) basis. We have

$$a(A_g) = ¼ (1 \times 4 \times 1 + 1 \times 0 \times 1 + 1 \times 0 \times 1 + 1 \times -4 \times 1) = 0$$

$$a(B_g) = ¼ (1 \times 4 \times 1 + 1 \times 0 \times -1 + 1 \times 0 \times 1 + 1 \times -4 \times -1) = 2$$

$$a(A_u) = ¼ (1 \times 4 \times 1 + 1 \times 0 \times 1 + 1 \times 0 \times -1 + 1 \times -4 \times -1) = 2$$

$$a(B_u) = ¼ (1 \times 4 \times 1 + 1 \times 0 \times -1 + 1 \times 0 \times -1 + 1 \times -4 \times 1) = 0$$

and the basis spans $2B_g + 2A_u$. We find the SALCs of each symmetry by applying the projection operator to each basis function $\phi_i = \Sigma_g \chi_k(g)\, g\, f_i$ in turn. For the SALCs of B_g symmetry, we obtain

$$\phi(p_1) = p_1 - p_4 - p_4 + p_1 = 2\ (p_1 - p_4)$$

$$\phi(p_2) = p_2 - p_3 - p_3 + p_2 = 2\ (p_2 - p_3)$$

$$\phi(p_3) = p_3 - p_2 - p_2 + p_3 = 2\ (p_3 - p_2)$$

$$\phi(p_4) = p_4 - p_1 - p_1 + p_4 = 2\ (p_4 - p_1)$$

giving the two normalised linearly independent combinations

$$\phi_1 \frac{1}{\sqrt{2}}\ (p_1 - p_4)$$

$$\phi_2 = \frac{1}{\sqrt{2}} (p_2 - p_3)$$

For the A_u SALCs we have

$$\phi(p_1) = p_1 + p_4 + p_4 + p_1 = 2\ (p_1 + p_4)$$

$$\phi(p_2) = p_2 + p_3 + p_3 + p_2 = 2\ (p_2 + p_3)$$

$$\phi(p_3) = p_3 + p_2 + p_2 + p_3 = 2\ (p_3 + p_2)$$

$$\phi(p_4) = p_4 + p_1 + p_1 + p_4 = 2\ (p_4 + p_1)$$

yielding the normalised combinations

$$\phi = \frac{1}{\sqrt{2}}(p_1 + p_4)$$

$$\phi_4 = \frac{1}{\sqrt{2}}(p_2 + p_3)$$

We therefore have two molecular orbitals of the form

$$\psi(B_g) = c_1\,\phi_1 + c_2\,\phi_2 = \frac{1}{\sqrt{2}}\,[c_1(p_1 - p_4) + c_2(p_2 - p_3)]$$

and two of the form

$$\psi(A_u) = c_3\,\phi_3 + c_4\,\phi_4 = \frac{1}{\sqrt{2}}\,[c_3(p_1 + p_4) + c_4(p_2 + p_3)]$$

We find the coefficients and orbital energies by solving the secular equations for each set of orbitals in turn.

For the B_g orbitals we have

$$\begin{pmatrix} H_{11} - ES_{11} & H_{12} - ES_{12} \\ H_{12} - ES_{12} & H_{22} - ES_{22} \end{pmatrix} \begin{pmatrix} c_1 \\ c_2 \end{pmatrix} = \begin{pmatrix} 0 \\ 0 \end{pmatrix}$$

The integrals are:

$$\begin{aligned} H_{11} &= \langle\phi_1|H|\phi_1\rangle \\ &= \tfrac{1}{2}\,\langle p_1 - p_4|H|p_1 - p_4\rangle \\ &= \tfrac{1}{2}\,(\langle p_1|H|p_1\rangle - \langle p_1|H|p_4\rangle - \langle p_4|H|p_1\rangle + \langle p_4|H|p_4\rangle \\ &= \tfrac{1}{2}\,(\alpha - 0 - 0 + \alpha) \\ &= \alpha \end{aligned}$$

$$H_{22} = <\phi_1|H|\phi_1>$$
$$= \tfrac{1}{2} <p_2 - p_3|H|p_2 - p_3>$$
$$= \tfrac{1}{2} (<p_2|H|p_2> - <p_2|H|p_3> - <p_3|H|p_2> + <p_3|H|p_3>$$
$$= \tfrac{1}{2} (\alpha - \beta - \beta + \alpha)$$
$$= \alpha - \beta$$

$$H_{12} = <\phi_1|H|\phi_2>$$
$$= \tfrac{1}{2} <p_1 - p_4|H|p_2 - p_3>$$
$$= \tfrac{1}{2} (<p_1|H|p_2> - <p_1|H|p_3> - <p_4|H|p_2>+<p_4|H|p_3>$$
$$= \tfrac{1}{2} (\beta - 0 - 0 + \beta)$$
$$= \beta$$

$$S_{12} = <\phi_1|\phi_2>$$
$$= \tfrac{1}{2} <p_1 - p_4|p_2 - p_3>$$
$$= \tfrac{1}{2} (<p_1|p_2> - <p_1|p_3> - <p_4|p_2> + <p_4|p_3>$$

giving

$$\begin{pmatrix} \alpha\text{-}E & \beta \\ \beta & \alpha\text{-}\beta\text{-}E \end{pmatrix} \begin{pmatrix} c_1 \\ c_2 \end{pmatrix} = \begin{pmatrix} 0 \\ 0 \end{pmatrix}$$

The secular determinant is

$$(\alpha\text{-}E)(\alpha\text{-}\beta\text{-}E) - \beta^2 = 0$$
$$E^2 + (\beta\text{-}2\alpha)E + (\alpha^2\text{-}\alpha\beta\text{-}\beta^2) = 0$$
$$E = \frac{-(\beta\text{-}2\alpha) \pm \sqrt{(\beta\text{-}2\alpha)^2 - 4(\alpha^2\text{-}\alpha\beta\text{-}\beta^2)}}{2}$$
$$= \frac{-\beta + 2\alpha \pm \sqrt{5}\beta}{2}$$
$$= \alpha + 0.6180\ \beta \text{ and } \alpha - 1.6180\ \beta$$

Substituting the first energy into the secular equations gives

$$- 06180\ \beta\ c_1 + \beta c_2 = 0$$

$$c_2 = 0.6180\ c_1$$

while the second energy gives

$$1.6180\ \beta\ c_1 + \beta c_2 = 0$$
$$c_2 = -1.6180\ c_1$$

Our two molecular orbitals of B_g symmetry are therefore

$$\begin{aligned}\psi_1(B_g) &= c_1(\phi_1 + 0.6180\ \phi_2)\\ &= 0.8506\ \phi_1 + 0.5257\ \phi_2\\ &= 0.6015\ p_1 + 0.3717\ p_2 - 0.3717\ p_3 - 0.6105\ p_4\end{aligned}$$

$$\begin{aligned}\psi_2(B_g) &= c_1(\phi_1 - 1.6180\ \phi_2)\\ &= 0.5257\ \phi_1 - 0.8506\ \phi_2\\ &= 0.3717\ p_1 - 0.6015\ p_2 + 0.6015\ p_3 - 0.3717\ p_4\end{aligned}$$

Now for the orbitals of A_u symmetry. The integrals we require are $H_{33} = \alpha$, $H_{44} = \alpha + \beta$, $H_{34} = \beta$, $S_{34} = 0$, giving the following secular equations

$$\begin{pmatrix}\alpha-E & \beta\\ \beta & \alpha+\beta-E\end{pmatrix}\begin{pmatrix}c_3\\ c_4\end{pmatrix} = \begin{pmatrix}0\\ 0\end{pmatrix}$$

Solving the secular determinant in the same way as above gives the energies

$$E = \frac{\beta + 2\alpha \pm \sqrt{5}\beta}{2} = \alpha + 1.6180\ \beta \text{ and } \alpha - 0.6180\ \beta$$

For the first energy, the secular equations give

$$-1.6180\ \beta\ c_1 + \beta c_2 = 0$$
$$c_2 = 1.6180\ c_1$$

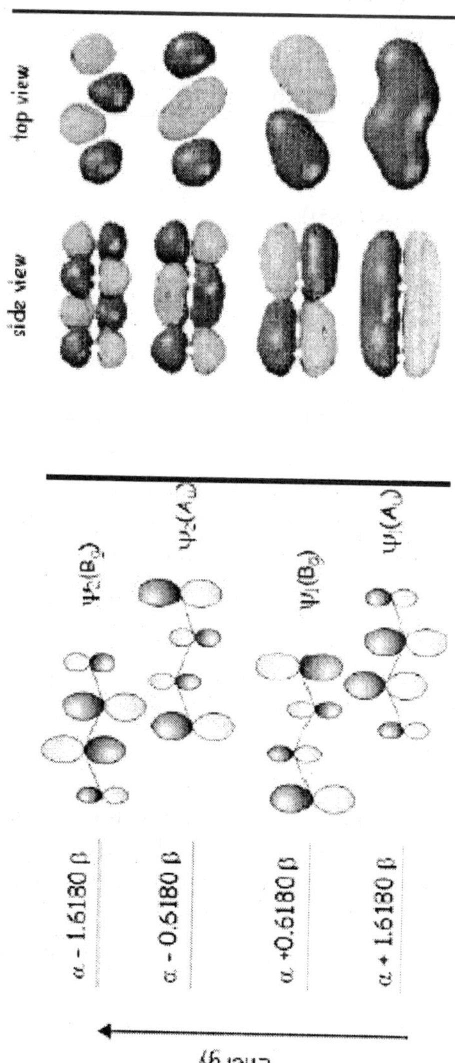

top view
side view
α − 1.6180 β
α − 0.6180 β
α + 0.6180 β
α + 1.6180 β
Energy

and for the second, we get

$$0.6180\ \beta c_1 + \beta c_2 = 0$$
$$c_2 = -0.6180\ c_1$$

The resulting molecular orbitals are

$$\begin{aligned}\psi_3(A_u) &= c_1(\phi_1 + 1.6180\ \phi_2)\\ &= 0.5257\ \phi_1 + 0.8506\ \phi_2\\ &= 0.3717\ p_1 + 0.6015\ p_2 + 0.6015\ p_3 + 0.3717\ p_4\end{aligned}$$

$$\begin{aligned}\psi_4(A_u) &= c_1(\phi_1 - 0.6180\ \phi_2)\\ &= 0.8506\ \phi_1 - 0.5257\ \phi_2\\ &= 0.6015\ p_1 - 0.3717\ p_2 - 0.\ 3717\ p_3 + 0.6015\ p_4\end{aligned}$$

The Huckel molecular orbitals for butadiene are shown below on the left. On the right are the orbitals obtained from much more advanced electronic structure calculations. We can see that Huckel theory does a fairly good job at predicting the qualitative form of the orbitals.

This course has given you a reasonable introduction to the qualitative description of molecular symmetry, and also to the way in which it can be used quantitatively within the context of group theory to predict important molecular properties. While this course has concentrated on the application of group theory to chemical bonding, there are many other areas in which it has proved extremely useful. For example, group theory may be used to predict the allowed vibrations of a polyatomic molecule, or to derive selection rules for spectroscopic transitions.

3

Vibrational Modes

C_{60} has been the focus of attention for different fields of science, after its preparation in solid form. From the physical point of view, relatively high T_c superconductivity and the five fold symmetry of the icosehedral C_{60} are of great importance. A detailed calculation of phonon density of state is necessary to understand the superconductivity mechanism in compounds of C_{60}.

On the basis of the force constant model one can calculate phonon density of state. The C_{60} cages consists of 20 hexagons and 12 pentagons arranged in a polyhedron known as 'truncated icosahedron'. In this case the total number of atoms is sixty and the symmetry group is icosahedral, I_h.

Group theoretical analysis gives 46 distinct vibrational modes, of which only 10 are raman active and 4 are infrared active. Both experimental and theoretical information on vibrational frequencies have been provided from inelastic neutron-scattering measurement. The existence of all 46 frequencies have also been provided from inelastic neutron-scattering measurements.

Several calculations of vibrational frequencies for C_{60} were reported by a number of authors. In many force constant models, only bond stretching force constants between nearest neighbour atoms have been considered, in addition to angle bending force constants. In such models many of the frequencies could not be reproduced with any reasonable accuracy, and different models which *t* the same data give widely different values for interatomic force constants.

In the present work, we propose a model that includes Keating parameters and force constants that connect further than nearest neighbour atoms for vibrational frequencies of isolated C_{60}. The effect of bonded and unbonded atoms to analyse the normal frequencies of C_{60}. One of the motivations for this work is to investigate the effect of the force constants between unbonded atoms in C_{60}.

In many models those were not taken into consideration, and consequently many normal frequencies did not appear. We present a group theoretical analyses to determine the types of the normal frequencies of C_{60}.

The site symmetry of icosahedral C_{60} and symmetry vector are also obtained in this section. Normal mode frequencies of C_{60} are calculated and listed by constructing a simple model. Finally, we discuss the effect of force constants between bonded and unbonded atoms and we compare our results with the experimental and some theoretical results.

Symmetry of C_{60}

C_{60} is the most important molecule which has icosehedral symmetry, with 120 symmetry operations. We begin with the standard analysis of molecular vibrations of C_{60} molecule. It is known that normal mode frequencies of C_{60} are qualitatively analysed by using symmetries associated with the corresponding irreducible representation. The group for C_{60} is the icosehedral group, I_h, and the generators which generate 120 symmetry operations of the icosehedral group are given by the matrices.

$$C_5 = \frac{1}{2}\begin{pmatrix} -\tau & -\sigma & 1 \\ \sigma & 1 & \tau \\ -1 & \tau & -\sigma \end{pmatrix}$$

$$C_2 = \begin{pmatrix} -1 & 0 & 0 \\ 0 & -1 & 0 \\ 0 & 0 & 1 \end{pmatrix} \quad I = -\begin{pmatrix} 1 & 0 & 0 \\ 0 & 1 & 0 \\ 0 & 0 & 1 \end{pmatrix} \tag{1}$$

where $\tau = ½ (1 + \sqrt{5})$ and $\sigma = ½ (-1 + \sqrt{5}$. The first matrix C_5 represents a rotation by $2\pi = 5$ about a fivefold axis. The second matrix C_2 is a cyclic permutation of the axes and the third matrix is the inversion. Generation relation of the generators are $(C_5)^5 = (C_2)^2 = (C_5C_2)^3 = I$. The 3 × 3 irreducible generators have been obtained by breaking the symmetries of SO(3) group. Table 1 which characterize C_{60} of full icosehedral group I_h is given below.

Table 1

Character table for the full icosehedral group I_h

I_h	E	$12C$	$12C$	$20C$	$15C$	I	$12S$	$12S$	$20S$	15σ	
A_g	1	1	1	1	1	1	1	1	1	1	
F_{1g}	3	τ	$-\sigma$	0	−1	$-\sigma$	τ	$-\sigma$	0	−1	
F_{2g}	3	$-\sigma$	τ	0	−1	3	τ	$-\sigma$	0	−1	R_x;
G_g	4	−1	−1	1	0	4	−1	−1	1	0	R_y;
H_g	5	0	0	−1	1	5	0	0	−1	1	R_z
A_u	1	1	1	1	1	−1	−1	−1	−1	−1	
F_{1u}	3	$-\tau$	$-\sigma$	0	−1	−3	$-\sigma$	$-\tau$	0	1	
F_{2u}	3	$-\sigma$	τ	0	−1	−3	$-\tau$	σ	0	1	
G_u	4	−1	−1	1	0	−4	1	1	−1	0	
H_u	5	0	0	−1	1	−5	0	0	1	−1	x y; z
χ_{site}	60	0	0	0	0	0	0	0	0	4	
G_{180}	180	0	0	0	0	0	0	0	0	4	

The site symmetry for an isolated C_{60} molecule can be found in the following simple way. We consider a radial vector r_i from the center of C_{60} molecule to ith atom in a site. The vectors associated with other atoms in a site are obtained by applying generators given in Eq. (1), on the vector r_i. After a few attempts, 60 × 60 reducible matrix generators, which transforms the radial vectors r_i to each other (i = 1, 2, . . . 60), are carried out yielding characters χ_{site} as given in Table 1. The symmetries of the electronic levels of the molecule can be found by taking the direct product of χ_{site} with A_g; F_{1u} and H_g for s; p and d functions, respectively.

This subject will be discussed elsewhere. The vibrational modes are qualitatively classified in a usual way such as direct product of 60 × 60 matrices, which are obtained from site symmetry with the generators given in Eq. (1), gives 180 × 180 matrix generators of the group I_h. Conjugacy classes 180 of 180 × 180 reducible representations is given in the Table 1. Decomposition of reducible representation 180 in terms of irreducible representation specifies the vibrational modes such as

$$\Gamma_{180} = 2A_g + 4F_{1g} + 4F_{2g} + 6G_g + 8H_g + A_u + 5F_{1u} + 5F_{2u} + 6G_u + 7H_u \quad (2)$$

The six degrees of freedom associated with the translations F_{1u} and rotations F_{1g} must be subtracted from 180 to classify the vibrational modes:

$$\Gamma_{vib} = 2A_g + 3F_{1g} + 4F_{2g} + 6G_g + 8H_g + A_u + 4F_{1u} + 5F_{2u} + 6G_u + 7H_u \quad (3)$$

We have seen that there are 174 degrees of freedom for isolated C_{60} molecule; a number of them are degenerate due to high symmetry. As such, 46 distinct mode frequencies are appearing for the C_{60} molecule in Eq. (3). Four fundamentals F_{1u} can appear in the infrared spectrum and ten $8H_g + 2A_g$ in the Raman spectrum.

Vibrational Modes

Some force constant models include only bondstretching force constants between two bonded atoms and bending force constants. But it is apparent that the interaction beyond nearest neighbours should be included to obtain the normal mode frequencies with a reasonable accuracy. In our approach, each atom in C_{60} is assumed to be a point mass, and they are connected with springs. The Keating type potential as employed for the C_{60} molecule can be written in the form

$$V = \sum_{i=1}^{N} \sum_{j=i+1}^{N} \frac{\alpha_{ij}}{2} [(u_i - u_j) \cdot \hat{r}_{ij}]^2 +$$

$$+ \frac{\beta}{2} \sum_{i=1}^{N} \sum_{j=i+1}^{N} \sum_{k=j+1}^{N} [(u_i - u_j) \cdot \hat{r}_{ik} + (u_k - u_i) \cdot \hat{r}_{jk}]^2, \quad (4)$$

where u_i is the small displacement of ith carbon atom about its equilibrium position, r_{ij} is the unit vector of r_i - r_j and i_j is the force constant between the ith and jth atoms in a C_{60} molecule can be considered as the angle bending force constant. In Eqn. (4) the summations are taken over all neighbour pairs, so that the value of N is 60. It is clear that the potential equation include 1770 different force constants α_{ij}, excluding angle bending force constant. Since the potential energy and kinetic energy of the C_{60} are separately invariant under symmetry operation, then we can write

$$R_{ij} V = V \quad (5)$$

where R_{ij} are the 180 × 180 reducible matrix generators of the Ih group. The potential energy can be simplified and number of force constants can be determined by the symmetry operation in Eq. (5). After performing a few calculations, we obtain only 23 different radial force constants of which two unbonded and the remaining 21 are unbonded force constants. Dynamical matrix $\Phi(\kappa\kappa')$ is obtained from the potential energy given in Eq. (4), such as

$$\Phi(\kappa\kappa') = \frac{\partial^2 V}{\partial u_i \partial u_j} \quad (i = 1, 2, \ldots 180) \text{ and } (j = 1, 2, \ldots 180). \quad (6)$$

Throughout this review, κ and κ' indicate the atom index and take on values for the number of atoms in a site. Eigenvalues of the dynamical matrix $\Phi(\kappa\kappa')$ of size 180 × 180 corresponds to the vibrational frequencies $m\omega^2$ of the C_{60}. The matrix $\Phi(\kappa\kappa')$ is block-diagonalized by using symmetry vectors. Eigenvalues of each block matrix gives vibrational eigenvalues for corresponding irreducible representation. The symmetry vectors belonging to the *i*th irreducible representation can be obtained by the projection method. In this work calculation of the symmetry vectors was carried out symbolically by mathematica, developing a simple method. The symmetry vectors for A_g and A_u representations are listed in Table 2.

We have carried out a calculation for eigenvalues of block diagonalized matrix $\Phi(\kappa\kappa')$, by considering interaction up to ten-neighbours.

So far, we have obtained dynamical matrix $\Phi(\kappa\kappa')$, that involves interactions among each atom and its ten neighbouring atoms. For comparison, we have also listed the 14 optically observed raman and infrared active modes, the modes measured by neutron inelastic scattering and high resolution electron-energy-loss spectroscopy, and some recent calculations. It can be shown that, with a few exceptions, all computed modes are

Table 2. Symmetry vectors associated to the irreducible representations A_2 and A_u

Γ	Symmetry Vector $(x_1, y_1, z_1; x_2, y_2, z_2 \quad \ldots \quad x_{60}, y_{60}, z_{60})$
	$\{0,0,4;\tau,-\sigma,-2;0,0,-4;\tau,-\sigma,2;-\tau,-\sigma,2;2,-\tau,\sigma;-\tau,-\sigma,-2;-\tau,\sigma,2;$ $-2,-\tau,-\sigma;-\tau,\sigma,-2;\tau,\sigma,-2;2,-\tau,-\sigma;\tau,-\sigma,2;-2,\tau,-\sigma;-2,-\tau,\sigma;-\sigma,-2,\tau;$ $2,\tau,\sigma;-\sigma,-2,-\tau;\sigma,-2,-\tau;-2,\tau,\sigma;\sigma,-2,\tau;-\sigma,2,-\tau;2,\tau,-\sigma;0,-4,0;$ $-\sigma,2,\tau;\sigma,2,\tau;0,-4,0;-4,0,0;-\sigma,2,-\tau;$ $-4,0,0;\sigma,-2,-\tau;4,0,0;0,4,0;-\sigma,-2,\tau;\sigma,-2,-\tau;0,4,0;-2,-\tau,-\sigma;$ $-\sigma,-2,-\tau;-\sigma,2,\tau;2,-\tau,\sigma;-\sigma,2\tau;\sigma,2,-\tau;-2,-\tau,\sigma;\sigma,2,\tau;2,\tau,\sigma;$ $2,-\tau,-\sigma;-\tau,-\sigma,2;-2,\tau,-\sigma;-\tau,-\sigma,-2;\tau,-\sigma,-2;2,\tau,-\sigma;2,\tau,-\sigma;\tau,-\sigma,2;$ $\tau,\sigma,-2;-2,\tau,\sigma;\tau,\sigma,2;0,0,-4;-\tau,\sigma,-2;0,0,4\}/8\sqrt{15}$
A_J	$\{0,-4,0;-2,-\tau,-\sigma;0,-4,0;-2,-\tau,\sigma;2,-\tau,\sigma;-\sigma,-2,-\tau;2,-\tau,-\sigma;-2,-\tau,-\sigma;$ $\sigma-2,\tau;-2,-\tau,\sigma;2,-\tau,\sigma;-\sigma,2\tau;2,-\tau,\sigma;-\sigma,-2,-\tau;\sigma-2,-\tau;-\tau,-\sigma,-2;$ $\sigma,-2,\tau;-\tau,-\sigma,2;\tau,-\sigma,2;-\sigma,-2,\tau;\tau,-\sigma,-2;-\tau,-\sigma,-2;\sigma,-2,-\tau;-4,0,0;\tau,-\sigma,2;$ $-\tau,-\sigma,2;4,0,0;0,0,-4;\tau,-\sigma,2;0,0,4;0,0,4;-\tau,\sigma,-2;$ $0,0,-4;-4,0,0;\tau,\sigma,2;-\tau,\sigma,2;4,0,0;-\sigma,2,-\tau;\tau,\sigma,-2;-\tau,\sigma,-2;\sigma,2,\tau;-\tau,\sigma,2;\tau,\sigma,2;$ $-\sigma,2,\tau;\tau,\sigma,-2;-\sigma,2,-\tau;\sigma,2,-\tau;-2,\tau,-\sigma;\sigma,2,\tau;-2,\tau,\sigma;-\sigma,2,\tau;$ $2,\tau,-\sigma;-2,\tau,-\sigma;\sigma,2,-\tau;-2,-\tau,\sigma;0,4,0;2,\tau,-\sigma;0,4,0\}/8\sqrt{15}$
A_u	$\{4,0,0;-\sigma,2,-\tau;-4,0,0;\sigma,-2,-\tau;\sigma,2,\tau;\tau,\sigma,-2;-\sigma,-2,\tau;\sigma,-2,\tau;-\tau,\sigma,2;$ $-\sigma,2,\tau;-\sigma,-2,-\tau;-\tau,-\sigma,-2;\sigma,2,-\tau;-\tau,-\sigma,2;\tau,-\sigma,2;2,-\tau,-\sigma;\tau,-\sigma;-2;$ $-2,-\tau,\sigma;\tau,\sigma,2;2,\tau;\sigma;-2,\tau,\sigma;-\tau,\sigma,-2;0,0,-4;2,\tau,-\sigma;2,-\tau,\sigma;$ $0,0,4;0,-4,0;-2,-\tau,-\sigma;0,4,0;0,-4,0;2,\tau,-\sigma;$ $0,4,0;0,0,4;-2,\tau,\sigma;-2,-\tau,-\sigma;0,0,-4;\tau,-\sigma,-2;2,-\tau,\sigma;$ $-2,-\tau,\sigma;-\tau,-\sigma,2;2,\tau,\sigma;2,-\tau,-\sigma;-\tau,\sigma,-2;-2,\tau,-\sigma;-\tau,\sigma,2;\tau,\sigma,2;-\sigma,$ $-2,-\tau;\tau,\sigma,-2;\sigma,-2,\tau;\tau,-\sigma,2;-\sigma,2,\tau;-\tau,-\sigma,-2;$ $-\sigma,-2,\tau;\sigma,2,-\tau;4,0,0;\sigma,-2,-\tau,\sigma;-4,0,0\}/8\sqrt{15}$

in agreement with experimental results. These comparisons give us confidence modes, and are in agreement with experimental results. These comparisons give us confidence that our model should be useful for interpretation of experimental data. This model should serve as the starting point for determination of the electron-phonon coupling strength and vibrational frequencies in doped C_{60}.

We also investigated the effect of the bond-stretching force constants α_1, α_2, α_3 and α_4 on the vibrational modes. It turns out that the A_u mode, the lowest A_g mode, the first lowest mode A_g mode, the first lowest F_{1g}, F_{2g}, F_{1u} and F_{2u} modes and the first lowest G_g and G_u; H_g and H_u modes not dependent on the values of α_1, α_2 and α_3. Therefore any force constant model should includes force between four or five shells of atoms. At least a large radial forth and fifth nearest neighbour force constant was needed whereas the model did not include a radial force constant between further and nearest neighbours. In fact, the force constants α_3 and α_4 should be nearly equal to each other.

4

Free Groups and Presentations

A group is a nonempty set G together with a law of composition $(a, b) \mapsto a * b : G \times G \to$ G satisfying the following axioms:

(a) (associative law) for all a, b, c e G,

$$(a * b) * c = a * (b * c);$$

(b) (existence of an identity element) there exists an element e ε G such that

$$a * a' = e * a$$

for all $a \varepsilon G$;

(c) (existence of inverses) for each $a \varepsilon G$, there exists an $a' \varepsilon$ G such that

$$a * a' = e = a' * a$$

When (a) and (b) hold, but not necessarily (c), we call (G, *) a semigroup. We usually abbreviate $(G,*)$ to G, and we usually write $a * b$ and e respectively as ab and 1, or as $a + b$ and 0. Two groups G and G' are isomorphic if there exists a one-to-one correspondence $a \leftrightarrow a'$, $G \leftrightarrow G'$, such that $(ab)' = a'b'$ for all $a, b \varepsilon G$.

In the following, $a, b, \ldots$ are elements of a group G.

(a) If $aa = a$, then $a = e$ (multiply by a' and apply the axioms). Thus e is the unique element of G with the property that $ee = e$.

(b) If $ba = e$ and $ac = e$, then

$$b = be = b(ac) = (ba)c = ec = c$$

Hence the element a' in (c) is uniquely determined by a. We call it the inverse of a, and denote it a^{-1} (or the negative of a, and denote it $-a$).

c) Note that (a) implies that the product of any ordered triple a_1, a_2, a_3 of elements of G is unambiguously defined: whether we form a_1a_2 first and then $(a_1a_2)a_3$, or a_2a_3 first and then $a_1(a_2a_3)$, the result is the same. In fact, (a) implies that the product of any ordered n-tuple $a_1, a_2, \ldots, a_n$ of elements of G is unambiguously defined. We prove this by induction on n. In one multiplication, we might end up with

$$(a_1 \cdots a_i)(a_{i+1} \cdots a_n) \tag{1}$$

as the final product, whereas in another we might end up with

$$(a_1 \cdots a_j)(a_{j+1} \cdots a_n). \tag{2}$$

Note that the expression within each pair of parentheses is well defined because of the induction hypotheses. Thus, if $i = j$, (1) equals (2). If $i \neq j$, we may suppose $i < j$. Then

$$(a_1 \cdots a_i)(a_{i+1} \cdots a_n) = (a_1 \cdots a_i)$$
$$((a_{i+1} \cdots a_j)(a_{j+1} \cdots a_n))$$

$$(a_1 \cdots a_j)(a_{j+1} \cdots a_n) =$$
$$((a_1 \cdots a_i)(a_{i+1} \cdots a_j))(a_{j+1} \cdots a_n)$$

and the expressions on the right are equal because of (a).

(d) The inverse of $a_1a_2 \cdots a_n$ is $a_n^{-1} a_{n-1}^{-1} \cdots a_1^{-1}$, i.e., the inverse of a product is the product of the inverses in the reverse order.

(e) Axiom (c) implies that the cancellation laws hold in groups:

$$ab = ac \Rightarrow b = c, \; ba = ca \Rightarrow b = c$$

(multiply on left or right by a^{-1}). Conversely, if G is finite, then the cancellation laws imply Axiom (c): the map $x \mapsto ax: G \to G$ is injective, and hence (by counting) bijective; in particular, 1 is in the image, and so a has a right inverse; similarly, it has a left inverse, and the argument in (b) above shows that the two inverses must then be equal.

The order of a group is the number of elements in the group. A finite group whose order is a power of a prime p is called a p-group.

For an element a of a group G, define

$$a^n = \begin{cases} aa \cdots a & n > 0 \quad (n \text{ copies of } a) \\ 1 & n = 0 \\ a^{-1}a^{-1} \cdots a^{-1} & n < 0 \quad (|n| \text{ copies of } a^{-1}) \end{cases}$$

The usual rules hold:

$$a^m a^n = a^{m+n}, \; (a^m)^n = a^{mn}. \tag{3}$$

It follows from (3) that the set

$$\{n \in Z \mid a^n = 1\}$$

is an ideal in Z. Therefore, this set equals (m) for some $m \geq 0$. When $m = 0$, a is said to have infinite order, and $a^n \neq 1$ unless $n = 0$. Otherwise, a is said to have finite order m, and m is the smallest integer > 0 such that $a^m = 1$; in this case, $a^n = 1 \Leftrightarrow m|n$; moreover $a^{-1} = a^{m-1}$.

Example 1.

(a) For $m \geq 1$, let $C_m = Z/mZ$, and for $m = \infty$, let $C_m = Z$ (regarded as groups under addition).

(b) Probably the most important groups are matrix groups. For example, let R be a commutative ring. If A is an $n \times n$ matrix with coefficients in R whose determinant is a unit in R, then the cofactor formula for the inverse of a matrix shows that A^{-1} also has coefficients in R. In more detail, if A' is the transpose of the matrix of cofactors of A, then $A \cdot A' = \det A \cdot I$, and so $(\det A)^{-1}A'$ is the inverse of A. It follows that the set $GL_n(R)$ of such matrices is a group. For example $GL_n(Z)$ is the group of all $n \times n$ matrices with integer coefficients and determinant ± 1. When R is finite, for example, a finite field, then $GL_n(_R)$ is a finite group. Note that $GL_1(R)$ is just the group of units in R — we denote it $R^{\times}$.

(c) If G and H are groups, then we can construct a new group $G \times H$, called the (direct) product of G and H. As a set, it is the cartesian product of G and H, and multiplication is defined by:

$$(g, h)(g', h') = (gg', hh').$$

(d) A group is commutative (or abelian) if

$$ab = ba, \text{ all } a, b \in G.$$

In a commutative group, the product of any finite (not necessarily ordered) set S of elements is defined.

Recall the classification of finite abelian groups. Every finite abelian group is a product of cyclic groups. If $gcd\ (m, n) = 1$, then $C_m \times C_n$ contains an element of order mn, and so $C_m \times C_n \approx C_{mn}$, and isomorphisms of this type give the only ambiguities in the decomposition of a group into a product of cyclic groups.

From this one finds that every finite abelian group is isomorphic to exactly one group of the following form:

$$C_{n1} \times \cdots \times C_{nr}, \; n_1|n_2, \ldots, n_{r-1}|n_r.$$

The order of this group is $n_1 \cdots n_r$.

For example, each abelian group of order 90 is isomorphic to exactly one of C_{90} or $C_3 \times C_{30}$ (note that n_r must be a factor of 90 divisible by all the prime factors of 90).

(e) *Permutation groups:* Let S be a set and let G be the set Sym(S) of bijections $\alpha: S \to S$. Then G becomes a group with the composition law $ab = \alpha \, o \, \beta$. For example, the permutation group on n letters is S_n = Sym({1, ..., n}), which has order n!. The symbol

$$\begin{pmatrix} 1 & 2 & 3 & 4 & 5 & 6 & 7 \\ 2 & 5 & 7 & 4 & 3 & 1 & 6 \end{pmatrix}$$

denotes the permutation sending $1 \mapsto 2$, $2 \mapsto 5$, $3 \mapsto 7$, etc..

Subgroups

Proposition 2. Let G be a group and let S be a nonempty subset of G such that

(a) $a, b \in S \Rightarrow ab \in S$;

(b) $a \in S \Rightarrow a^{-1} \in S$.

Then the law of composition on G makes S into a group.

Therefore,

$$A \cdot (A^{n-1} + a_{n-1}A^{n-2} + \cdots) \mp (\det A) \cdot I,$$

and

$$A \cdot (A^{n-1} + a_{n-1}A^{n-2} + \cdots) \cdot \mp (\det A)^{-1} = I,$$

Proof. Condition (a) implies that the law of composition on G does define a law of composition $S \times S \rightarrow S$ on S, which is automatically associative. By assumption S contains at least one element a, its inverse a^{-1}, and the product $1 = aa^{-1}$. Finally (b) shows that inverses exist in S.

A subset S as in the proposition is called a subgroup of G.

If S is finite, then condition (a) implies (b): let $a \in S$; then $\{a, a^2, \ldots\} \subset S$, and so a has finite order, say $a^n = 1$; now $a^{-1} = a^{n-1} \in S$. The example $(N, +) \subset (Z, +)$ shows that (a) does not imply (b) when S is infinite.

Proposition 3. An intersection of subgroups of G is a subgroup of G

Proof. It is nonempty because it contains 1, and conditions (a) and (b) of (2) are obvious.

Remark 4. It is generally true that an intersection of subobjects of an algebraic object is a subobject. For example, an intersection of subrings is a subring, an intersection of submodules is a submodule, and so on.

*Proposition 5.*For any subset X of a group G, there is a smallest subgroup of G containing X. It consists of all finite products of elements of X and their inverses.

Proof. The intersection S of all subgroups of G containing X is agains a subgroup containing X, and it is evidently the smallest such group. Clearly S contains with X, all finite prodcuts of elements of X and their inverses. But the set of such products satisfies (a) and (b) of (3) and hence is a subgroup containing X. It therefore equals S.

We write $\langle X \rangle$ for the subgroup S in the proposition, and call it the subgroup generated by X. For example, $\langle \theta \rangle = \{1\}$. If every element of G has finite order, for example, if G is finite, then the set of all finite products of elements of X is already a group (recall that if $a^m = 1$, then $a^{-1} = a^{m-1}$) and so equals $\langle X \rangle$.

We say that X generates G if $G = \langle X \rangle$, i.e., if every element of G can be written as a finite product of elements from X and their inverses. Note that the order of an element a of a group is the order of the subgroup $\langle a \rangle$ it generates.

Example 6.

(a) A group is cyclic if it is generated by one element, i.e., if $G = \langle \sigma \rangle$ for some $\sigma \in G$. If σ has finite order n, then

$$G = \{1, \sigma, \sigma^2, ..., \sigma^{n-1}\} \approx C_n, \quad \sigma^i \leftrightarrow i \bmod n,$$

and G can be thought of as the group of rotational symmetries (about the centre) of a regular polygon with n-sides. If σ has infinite order, then

$$G = \{. . . , \sigma^{-i}, . . . , \sigma^{-1}, 1, \sigma, . . . , \sigma^i, . . .\} \approx C_\infty, \quad \sigma^i \leftrightarrow i.$$

In future, we shall (loosely) use C_m to denote any cyclic group of order m (not necessarily Z/mZ or Z).

(b) Dihedral group, D_n.[6] This is the group of symmetries of a regular polygon with nsides. Number the vertices 1, . .. , n in the counterclockwise direction. Let σ be the rotation through $2\pi/n$ (so $i \to i + 1 \bmod n$), and let τ be the rotation (=reflection) about the axis of symmetry through 1 and the centre of the polygon (so $i \to n + 2 - i \bmod n$). Then

$$\sigma^n = 1; \ \tau^2 = 1; \ \tau\sigma\tau^{-1} = \sigma^{-1} \text{ (or } \tau\sigma = \sigma^n\text{-}1\tau).$$

The group has order $2n$; in fact

$$D_n = \{1, \sigma, ..., \sigma^{n-1}, \tau, ..., \sigma^{n-1}\tau\}.$$

(c) Quaternion group Q: Let $a =$

$$\begin{pmatrix} 0 & \sqrt{-1} \\ \sqrt{-1} & 0 \end{pmatrix}, b = \begin{pmatrix} 0 & 1 \\ -1 & 0 \end{pmatrix}$$

Then

$$a^4 = 1,\ a^2 = b^2,\ bab^{-1} = a^{-1}.$$

The subgroup of $GL_2(C)$ generated by a and b is

$$Q = \{1,\ a,\ a^2,\ a^3,\ b,\ ab,\ a^2b,\ a^3b\}.$$

The group Q can also be described as the subset $\{\pm 1, \pm i, \pm j, \pm k\}$ of the quaternion algebra.

(d) Recall that Sn is the permutation group on $\{1, 2, ..., n\}$. The alternating group A_n is the subgroup of even permutations. It has order $n!/2$.

Groups of Small Order

Every group of order <16 is isomorphic to exactly one on the following list:

1: C_1. 2: C_2. 3: C_3

4: C_4, $C_2 \times C_2$ (Viergruppe; Klein 4-group)

5: C_5

6: C_6, $S_3 = D_3$ (S_3 is the first noncommutative group)

7: C_7

8: C_8, $C_2 \times C_4$, $C_2 \times C_2 \times C_2$, Q, D_4

9: C_9, $C_3 \times C_3$

10: C_{10}, D_5

11: C_{11}

12: C_{12}, $C_2 \times C_6$, $C_2 \times S_3$, A_4, $C_3 \times C_4$

13: C_{13}

14: C_{14}, D_7

15: C_{15}

16: (14 groups)

For each prime p, there is only one group (up to isomorphism), namely C^p, and only two groups of order p^2, namely, $C_p \times C_p$ and C_{p2}. Roughly speaking, the more high powers of primes divide n, the more groups of order n you expect. In fact, if $f(n)$ is the number of isomorphism classes of groups of order n, then

$$f(n) \leq n^{(2/27+o(1))e(n)2}$$

where $e(n)$ is the largest exponent of a prime dividing n and $\sigma(1) \to 0$ as $e(n) \to \infty$.

By 2001, a complete irredundant list of groups of order ≤ 2000 had been found — up to isomorphism, there are 49, 910, 529, 484, 1–4 (electronic)).

Multiplication Tables

A law of composition on a finite set can be described by its multiplication table:

	1	a	b	c	...
1	1	a	b	c	...
a	a	a^2	ab	ac	...
b	b	ba	b^2	bc	...
c	c	ca	cb	c^2	...
⋮	⋮	⋮	⋮	⋮	

If the law of composition defines a group, then, because of the cancellation laws, each row (and each column) is a permutation of the elements of the group. This suggests an algorithm for finding all groups of a given finite order n, namely, list all possible multiplication tables and check the axioms. Except for very small n, this is not practical! The table has n^2 positions, and if we allow each position to hold any of the n elements, that gives a total of n^{n2} possible tables. Note how few groups there are. The 8^{64} = 6277 101 735 386 680 763 835 789 423 207 666 416 102 355 444 464 034 512 896 possible multiplication tables for a set with 8 elements give only 5 isomorphism classes of groups.

Homomorphisms

A homomorphism from a group G to a second G' is a map $\alpha : G \to G'$ such that $\alpha(ab) = \alpha(a)\,\alpha(b)$ for all $a, b \in G$.

Note that an isomorphism is simply a bijective homomorphism.

Let α be a homomorphism. Then

$$\alpha(a^m) = \alpha(a^{m-1} \cdot a) = \alpha(a^{m-1}) \cdot \alpha(a),$$

and so, by induction, $\alpha(a^m) = \alpha(a)^m$, $m \geq 1$. Moreover $\alpha(1) = \alpha(1 \times 1) = \alpha(1)\,\alpha(1)$, and so $\alpha(1) = 1$. Also $aa^{-1} = 1 = a^{-1}a \Rightarrow \alpha(a)\,\alpha(a^{-1}) = 1 = \alpha(a^{-1})\,\alpha(a)$, and so $\alpha(a^{-1})\ (a)^{-1}$. From this it follows that

$$\alpha(a^m) = \alpha(a)^m \text{ all } m \in Z.$$

We saw above that each row of the multiplication table of a group is a permutation of the elements of the group. As Cayley pointed out, this allows one to realize the group as a group of permutations.

Theorem (Cayley). There is a canonical injective homomorphism

$$\alpha : G \to \mathrm{Sym}(G).$$

For a 2 $\in G$, define $a_L : G \to G$ to be the map $x \mapsto ax$ (left multiplication by a).

For $x \in G$,

$$(a_L \circ b_L)(x) = a_L(b_L(x)) = a_L(bx) = abx = (ab)_L(x),$$

and so $(ab)_L = a_L \circ b_L$. As $1_L = id$, this implies that

$$a_L \circ (a^{-1})_L = id = (a^{-1})L \circ a_L,$$

and so aL is a bijection, i.e., $a_L \varepsilon$ Sym(G). Hence $a \mapsto a_L$ is a homomorphism $G \to$ Sym(G), and it is injective because of the cancellation law.

Corollary: A finite group of order n *can be identified with a subgroup of* S_n.

Proof: Number the elements of the group $a_1, \ldots, a_n$.

Unfortunately, when G has large order n, Sn is too large to be manageable. We shall see later that G can often be embedded in a permutation group of much smaller order than n!.

Cosets

Let H be a subgroup of G. A *left coset* of H in G is a set of the form

$$aH = \{ah \mid h \varepsilon H\},$$

some fixed $a \varepsilon G$; a *right coset* is a set of the form

$$Ha = \{ha \mid h \varepsilon H\},$$

some fixed $a \varepsilon G$.

Example: Let $G = R^2$, regarded as a group under addition, and let H be a subspace of dimension 1 (line through the origin). Then the cosets (left or right) of H are the lines parallel to H.

Proposition:

(a) If C is a left coset of H, and a εC, then $C = aH$.

(b) Two left cosets are either disjoint or equal.

(c) $aH = bH$ if and only if $a^{-1} b \varepsilon H$.

(d) Any two left cosets have the same number of elements (possibly infinite).

Proof:

(a) Because C is a left coset, $C = bH$ some $b \varepsilon G$, and because $a \varepsilon C$, $a = bh$ for some $h \varepsilon H$. Now $b = ah^{-1} \varepsilon aH$, and for any other element c of C, $c = bh^{0} = ah^{-1}h' \varepsilon aH$. Thus, $C \subset aH$. Conversely, if $c \varepsilon aH$, then $c = ah' = bhh' \varepsilon bH$.

(b) If C and C' are not disjoint, then there is an element $a \varepsilon C \cap C'$, and $C = aH$ and $C' = aH$.

(c) We have $aH = bH \Leftrightarrow$ b $\varepsilon aH \Leftrightarrow b = ah$, for some $h \varepsilon H$, i.e., $\Leftrightarrow a^{-1} b \varepsilon H$.

(d) The map $(ba^{-1})_L : ah \mapsto bh$ is a bijection $aH \to bH$.

Particularly, the left cosets of H in G partition G, and the condition "a and b lie in the same left coset" is an equivalence relation on G.

The index $(G : H)$ of H in G is defined to be the number of left cosets of H in G. In particular, $(G : 1)$ is the order of G. Each left coset of H has $(H: 1)$ elements and G is a disjoint union of the left cosets.

When G is finite, we can conclude:

Theorem: If G is finite, then

$$(G : 1) = (G : H)(H : 1).$$

In particular, the order of H divides the order of G.

Corollary: The order of every element of a finite group divides the order of the group.

Proof: Apply Lagrange's theorem to $H = \langle g \rangle$, recalling that $(H : 1) =$ order (g).

Example: If G has order p, a prime, then every element of G has order 1 or p. But only e has order 1, and so G is generated by any element $g \neq e$. In particular, G is cyclic,

$G \approx C_p$. Hence, up to isomorphism, there is only one group of order 1, 000, 000, 007; in fact there are only two groups of order 1, 000, 000, 014, 000, 000, 049.

Remark:

(a) There is a one-to-one correspondence between the set of left cosets and the set of right cosets, viz, aH ® Ha^{-1}. Hence $(G : H)$ is also the number of right cosets of H in G. But, in general, a left coset will not be a right coset.

(b) Lagrange's theorem has a partial converse: if a prime p divides $m = (G : 1)$, then G has an element of order p; if pn divides m, then G has a subgroup of order p_n. However, note that $C_2 \times C_2$ has order 4, but has no element of order 4, and $A4$ has order 12, but it has no subgroup of order 6. More generally, we have the following result.

Proposition: Let G be a finite group. If $G \supset H \supset K$ with H and K subgroups of G, then

$$(G : K) = (G : H)\ (H : K).$$

Proof: Write $G = \cup_{gi} H$ (disjoint union), and $H = \cup h_j K$ (disjoint union). On multiplying the second equality by g_i, we find that $g_i H = \cup_j g_i h_j K$ (disjoint union), and so $G = \cup g_i h_j K$ (disjoint union).

Normal Subgroups

When S and T are two subsets of a group G, we let

$$ST = \{st \mid s \in S,\ t \in T\}.$$

A subgroup N of G is normal, written $N\ G$, if $gNg^{-1} = N$ for all $g \in G$. An intersection of normal subgroups of a group is normal.

To show N normal, it suffices to check that $gNg^{-1} \subset N$ for all g, because $gNg^{-1} \subset N \Rightarrow g^{-1}gNg^{-1}g \Rightarrow g^{-1}Ng$ (multiply left and right with g^{-1} and g); hence $N \subset g^{-1}Ng$ for all g, and, on rewriting this with g^{-1} for g, we find that $N \subset gNg^{-1}$ for all g.

The next example shows however that there can exist an N and a g such that $gNg^{-1} \subset N$, $gNg^{-1} \neq N$.

Example: Let $G = GL_2(Q)$, and let

$$H = \{\left(\begin{smallmatrix} 1 & n \\ 0 & 1 \end{smallmatrix}\right) \mid n \in \mathbb{Z}\}.$$

Then H is a subgroup of G; in fact it is isomorphic to Z. Let

$$g = \begin{pmatrix} 0 & 5 \\ 0 & 1 \end{pmatrix}$$

Then

$$g \begin{pmatrix} 1 & n \\ 0 & 1 \end{pmatrix} g^{-1} =$$

$$\begin{pmatrix} 5 & 5n \\ 0 & 1 \end{pmatrix} \begin{pmatrix} 5^{-1} & 0 \\ 0 & 1 \end{pmatrix} = \begin{pmatrix} 1 & 5n \\ 0 & 1 \end{pmatrix}$$

Proposition: A subgroup N of G is normal if and only if each left coset of N in G is also a right coset, in which case, $gN = Ng$ for all $g \in G$.

Proof: $\Rightarrow$ Multiply the equality $gNg^{-1} = N$ on the right by g.

$\Leftarrow$ (: If gN is a right coset, then it must be the right coset Ng. Hence $gN = Ng$, and so $gNg^{-1} = N$. This holds for all g.

Remark: In other words, in order for N to be normal, we must have that for all $g \in G$ and $n \in N$, there exists an $n' \in N$ such that $gn = n'g$ (equivalently, for all $g \in G$ and $n \in N$, there exists an n ε such that $ng = gn'$.) Thus, an element of G can be moved past an element of N at the cost of replacing the element of N by a different element of N.

Example: (a) Every subgroup of index two is normal. Indeed, let $g \in G$, $g \notin H$. Then $G = H \cup gH$ (disjoint union). Hence gH is the complement of H in G. The same argument shows that Hg is the complement of H in G. Hence $gH = Hg$.

(b) Consider the dihedral group $D_n = \{1, \sigma, \ldots, \sigma^{n-1}, \tau, \ldots, \sigma^{n-1}\tau\}$. Then $C_n = \{1, \sigma, \ldots, \sigma^{n-1}\}$ has index 2, and hence is normal. For $n \geq 3$ the subgroup $\{1, \tau\}$ is not normal because $\sigma\tau\sigma^{-1} = \tau\sigma^{n-2} \notin \{1, \tau\}$.

(c) Every subgroup of a commutative group is normal (obviously), but the converse is false: the quaternion group Q is not commutative, but every subgroup is normal.

A group G is said to be *simple* if it has no normal subgroups other than G and $\{1\}$. Such a group can have still lots of nonnormal subgroups — in fact, the Sylow theorems imply that every group has nontrivial subgroups unless it is cyclic of prime order.

Proposition: If H and N are subgroups of G and N is normal, then

$$HN = \{hn \mid h \in H, n \in N\}$$

is a subgroup of G. If H is also normal, then HN is a normal subgroup of G.

Proof: The set HN is nonempty, and

$$(hn)(h'n') = hh'n''n' \in HN,$$

and so it is closed under multiplication. Since

$$(hn)^{-1} = n^{-1}h^{-1} = h^{-1}n' \in HN$$

it is also closed under the formation of inverses. If both H and N are normal, then

$$gHNg^{-1} = gHg^{-1} \cdot gNg^{-1} = HN$$

for all $g \in G$.

Quotients

The kernel of a homomorphism $\alpha\ G \to G'$ is

$$\text{Ker}(\alpha) = \{g \in G | \alpha\ (g) = 1\}$$

If is injective, then Ker(α) = {1}. Conversely, if Ker(α) = 1 then α is injective, because

$$\alpha(g) = \alpha(g')\ ) \Rightarrow \alpha\ (g^{-1}g') = 1 \Rightarrow g^{-1}g' = 1 \Rightarrow g = g'.$$

Proposition: The kernel of a homomorphism is a normal subgroup.

Proof: It is obviously a subgroup, and if $a\ \alpha$ Ker(α) so that $\alpha(a) = 1$, and $g \in G$, then

$$\alpha(gag^{-1}) = \alpha(g)\ \alpha(a)\ \alpha(g)^{-1} = \alpha(g)\ \alpha(g)^{-1} = 1.$$

Hence $gag^{-1} \in$ Ker α.

Proposition: Every normal subgroup occurs as the kernal of a homomorphism. More precisely, if N is a normal subgroup of G, then there is a natural group structure on the set of cosets of N in G.

Proof: Write the cosets as left cosets, and define $(aN)(bN) = (ab)N$. We have to check (a) that this is well-defined, and (b) that it gives a group structure on the set of cosets. It will then be obvious that the map $g \mapsto gN$ is a homomorphism with kernel N.

Check (a). Suppose $aN = a'N$ and $bN = b'N$; we have to show that $abN = a'b'N$. But we are given that $a = a'n$ and $b = b'n'$, some $n, n' \in N$. Hence

$$ab = a'nb'n' = a'b'n''n' \in a'b'N.$$

Therefore abN and $a'b'N$ have a common element, and so must be equal. Checking (b) is straightforward: the set is nonempty;

the associative law holds; the coset N is an identity element; $a^{-1}N$ is an inverse of aN.

When N is a normal subgroup, we write G/N for the set of left (= right) cosets of N in G, regarded as a group. It is called the quotient of G by N. The map $a \mapsto aN : G \to G/N$ is a surjective homomorphism with kernel N. It has the following universal property: for any homomorphism $\alpha : G \to G'$ of groups such that $\alpha(N) = 1$, there exists a unique homomorphism $G/N \to G'$ such that the following diagram commutes:

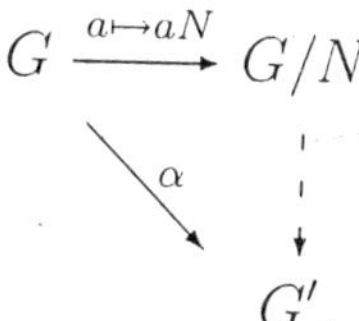

Example: (a) Consider the subgroup mZ of Z. The quotient group Z/mZ is a cyclic group of order m.

(b) Let L be a line through the origin in R^2. Then R^2/L is isomorphic to R (because it is a one-dimensional vector space over R).

(c) The quotient $D_n/\langle\sigma\rangle \approx \{1, \tau\}$ (cyclic group of order 2).

1. Show that the quaternion group has only one element of order 2, and that it commutes with all elements of Q. Deduce that Q is not isomorphic to D_4, and that every subgroup of Q is normal.

2. Consider the elements

$$a = \begin{pmatrix} 0 & -1 \\ 1 & 0 \end{pmatrix} \quad b = \begin{pmatrix} 0 & 1 \\ -1 & -1 \end{pmatrix}$$

in $GL_2(Z)$. Show that $a^4 = 1$ and $b^3 = 1$, but that ab has infinite order, and hence that the group $\langle a, b\rangle$, is infinite.

3. Show that every finite group of even order contains an element of order 2.
4. Let N be a normal subgroup of G of index n. Show that if $g \in G$, then $g^n \in N$. Give an example to show that this may be false when N is not normal.

Free Groups and Presentations

It is frequently useful to describe a group by giving a set of generators for the group and a set of relations for the generators from which every other relation in the group can be deduced. For example, D_n can be described as the group with generators σ, τ and relations

$$\sigma^n = 1, \ \tau^2 = 1, \ \tau\sigma\tau\sigma = 1.$$

In this section, we make precise what this means. First we need to define the free group on a set X of generators — this is a group generated by X and with no relations except for those implied by the group axioms. Because inverses cause problems, we first do this for semigroups.

Free Semigroups

Recall that (for us) a semigroup is a set G with an associative law of composition having an identity element 1. A homomorphism : $\alpha : S \to S'$ of semigroups is a map such that $\alpha(ab)\ \alpha(a)\alpha(b)$ for all $a, b \in S$ and $\alpha\ (1) = 1$. Then σ preserves all finite products.

Let $X = \{a, b, c, \ldots\}$ be a (possibly infinite) set of symbols. A word is a finite sequence of symbols in which repetition is allowed. For example,

$$aa, \ aabac, \ b$$

are distinct words. Two words can be multiplied by juxtaposition, for example,

$$aaaa\ aabac = aaaaaabac.$$

This defines on the set W of all words an associative law of composition. The empty sequence is allowed, and we denote it by 1. (In the unfortunate case that the symbol 1 is already an element of X, we denote it by a different symbol.) Then 1 serves as an identity element. Write SX for the set of words together with this law of composition. Then SX is a semigroup, called the *free semigroup* on X. When we identify an element a of X with the word a, X becomes a subset of SX and generates it (i.e., there is no proper subsemigroup of SX containing X). Moreover, the map $X \to SX$ has the following universal property: for any map (of sets) $\alpha : X \to S$ from X to a semigroup S, there exists a unique homomorphism $SX \to S$ making the following diagram commute:

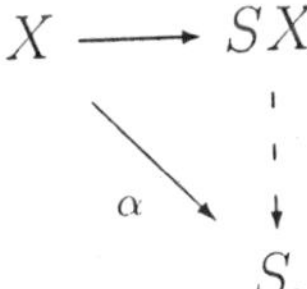

In fact, the unique extension of α takes the value:

$$\alpha\ (1) = 1_S,\ \alpha(dba \cdots) = \alpha(d)\alpha(b)(a) \cdots .$$

Free Groups

We want to construct a group FX containing X and having the same universal property as SX with "semigroup" replaced by "group". Define X' to be the set consisting of the symbols in X and also one additional symbol, denoted a^{-1}, for each $a \in X$; thus

$$X' = \{a, a^{-1}, b, b^{-1}, \ldots\}.$$

Let W' be the set of words using symbols from X'. This becomes a semigroup under juxtaposition, but it is not a group because we cannot cancel out the obvious terms in words of the following form:

$$\cdots xx^{-1} \cdots \text{ or } \cdots x^{-1}x \cdots$$

A word is said to be reduced if it contains no pairs of the form xx^{-1} or $x^{-1}x$. Starting with a word w, we can perform a finite sequence of cancellations to arrive at a reduced word (possibly empty), which will be called the reduced form of w. There may be many different ways of performing the cancellations, for example,

$$cabb^{-1}a^{-1}c^{-1}ca \mapsto caa^{-1}c^{-1}ca \mapsto cc^{-1}ca \mapsto ca \ :$$

$$cabb^{-1}a^{-1}c^{-1}ca \mapsto cabb^{-1}a^{-1}a \mapsto cabb^{-1} \mapsto ca.$$

We have underlined the pair we are cancelling. Note that the middle a^{-1} is cancelled with different a's, and that different terms survive in the two cases. Nevertheless we ended up with the same answer, and the next result says that this always happens.

Proposition: There is only one reduced form of a word.

Proof: We use induction on the length of the word w. If w is reduced, there is nothing to prove. Otherwise a pair of the form xx^{-1} or $x^{-1}x$ occurs — assume the first, since the argument is the same in both cases. It is observed that any two reduced forms of w obtained by a sequence of cancellations in which xx^{-1} is cancelled first are equal, because the induction hypothesis can be applied to the (shorter) word obtained by cancelling xx^{-1}.

Next observe that any two reduced forms of w obtained by a sequence of cancellations in which xx^{-1} is cancelled at some point are equal, because the result of such a sequence of cancellations will not be affected if xx^{-1} is cancelled first. Finally, consider a reduced form w' obtained by a sequence in which no cancellation cancels xx^{-1} directly. Since xx^{-1} does not remain in

w', at least one of x or x^{-1} must be cancelled at some point. If the pair itself is not cancelled, then the first cancellation involving the pair must look like

$$\cdots x^{-1}\, xx^{-1} \cdots \text{ or } \cdots x\, x^{-1}\, x \cdots$$

where our original pair is underlined. But the word obtained after this cancellation is the same as if our original pair were cancelled, and so we may cancel the original pair instead. Thus we are back in the case just proved. We say two words w, w' are *equivalent,* denoted $w \sim w'$, if they have the same reduced form. This is an equivalence relation.

Proposition: Products of equivalent words are equivalent, i.e.,

$$w \sim w',\ v \sim v' \Rightarrow wv\ w'v'.$$

Proof: Let w' and v' be the reduced forms of w and of v. To obtain the reduced form of wv, we can first cancel as much as possible in w and v separately, to obtain $w'v'$ and then continue cancelling. Thus the reduced form of wv is the reduced form of $w'v'$. A similar statement holds for $w'v'$, but (by assumption) the reduced forms of w and v equal the reduced forms of w' and v', and so we obtain the same result in the two cases.

Let FX be the set of equivalence classes of words. The proposition shows that the law of composition on W_0 defines a law of composition on FX, which obviously makes it into a semigroup. It also has inverses, because

$$ab \cdots gh \cdot h^{-1}g^{-1} \cdots b^{-1}a^{-1} \sim 1.$$

Thus FX is a group, called the *free group* on X. To summarize: the elements of FX are represented by words in X'; two words represent the same element of FX if and only if they have the same reduced forms; multiplication is defined by juxtaposition; the empty word represents 1; inverses are obtained in the obvious way. Alternatively, each element of FX is represented by a unique reduced word; multiplication is defined by juxtaposition and passage to the reduced form.

When we identify $a \varepsilon X$ with the equivalence class of the (reduced) word a, then X becomes identified with a subset of FX — clearly it generates FX. The next proposition is a precise statement of the fact that there are no relations among the elements of X when regarded as elements of FX except those imposed by the group axioms.

Proposition: For any map (of sets) $X \to G$ from X to a group G, there exists a unique homomorphism $FX \to G$ making the following diagram commute:

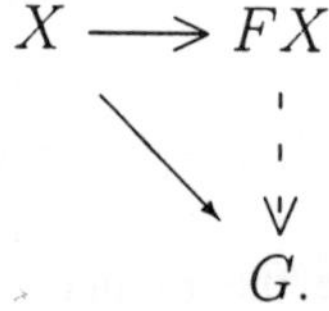

Proof: Consider a map $\alpha : X \to G$. We extend it to a map of sets $X' \to G$ by setting $\alpha(a^{-1}) = (a)^{-1}$. Because G is, in particular, a semigroup, extends to a homomorphism of semigroups $SX' \to G$. This map will send equivalent words to the same element of G, and so will factor through $FX = SX'{\sim}$. The resulting map $FX \to G$ is a group homomorphism. It is unique because we know it on a set of generators for FX.

The universal property of the map $\iota : X \to FX$, $x \mapsto x$, characterizes it: if $\iota : X \to F'$ is a second map with the same universal property, then there is a unique isomorphism $\alpha : FX \to F'$ such that $\to (\iota x) = \iota' x$ for all $x \varepsilon X$.

Corollary: Every group is a quotient of a free group.

Proof: Choose a set X of generators for G (e.g., $X = G$), and let F be the free group generated by X. According to (2.3), the inclusion $X \to G$ extends to a homomorphism $F \to G$, and the image, being a subgroup containing X, must equal G.

The free group on the set $X = \{a\}$ is simply the infinite cyclic group C_∞ generated by a, but the free group on a set consisting of two elements is already very complicated. Without proof, some important results on free groups.

Theorem: (Nielsen-Schreier) Subgroups of free groups are free.

The best proof uses topology, and in particular covering spaces.

Two free groups FX and FY are isomorphic if and only if X and Y have the same number of elements. Thus we can define the rank of a free group G to be the number of elements in (i.e., cardinality of) a free generating set, i.e., subset $X \subset G$ such that the homomorphism $FX \to G$ given by is an isomorphism. Let H be a finitely generated subgroup of a free group F. Then there is an algorithm for constructing from any finite set of generators for H a free finite set of generators. If F has rank n and $(F : H) = i < \infty$, then H is free of rank

$$ni - i + 1.$$

In particular, H may have rank greater than that of F.

Generators and Relations

An intersection of normal subgroups is again a normal subgroup. Therefore, just as for subgroups, we can define the normal subgroup generated by a set S in a group G to be the intersection of the normal subgroups containing S. Its description in terms of S is a little complicated. Call a subset S of a group G normal if $gSg^{-1} \subset S$ for all $g \in G$. Then it is easy to show:

(a) if S is normal, then the subgroup $\langle S \rangle$ generated by it is normal;

(b) for $S \subset G$, $\cup_{g \in G} gSg^{-1}$ is normal, and it is the smallest normal set containing S. From these observations, it follows that:

Lemma: The normal subgroup generated by S $\subset$ G is $\langle_{g\in G}\, gSg^{-1}\rangle$.

Consider a set X and a set R of words made up of symbols in X'. Each element of R represents an element of the free group FX, and the quotient G of FX by the normal subgroup generated by these elements is said to have X as *generators* and R as *relations*. One also says that (X, R) is a *presentation* for G, $G = \langle X|R\rangle$, and that R is a set of defining relations for G.

Example: (a) The dihedral group D_n has generators σ, τ and defining relations

$$\sigma^n,\ \tau^2,\ \tau\sigma\tau\sigma$$

(b) The generalized quaternion group Q_n, $n \geq 3$, has generators a, b and relations $a^{2n-1} = 1$, $a^{2n-2} = b^2$, $bab^{-1} = a^{-1}$. For $n = 3$ this is the group Q of. In general, it has order 2^n.

(c) Two elements a and b in a group commute if and only if their commutator $[a, b] =_{df} aba^{-1}b^{-1}$ is 1. The free abelian group on generators $a_1, \ldots, a_n$ has generators $a_1, a_2, \ldots,$ a_n has generators $a_1, a_2 \ldots\ldots a_n$ and relations

$$[a_i, a_j],\ i \neq j.$$

For the remaining examples, which contains a good account of the interplay between group theory and topology. For example, for many types of topological spaces, there is an algorithm for obtaining a presentation for the fundamental group.

(d) The fundamental group of the open disk with one point removed is the free group on σ where σ is any loop around the point

(e) The fundamental group of the sphere with r points removed has generators $\sigma_1, \ldots, \sigma_r$ (σ_i is a loop around the ith point) and a single relation

$$\sigma_1 \cdots \sigma_r = 1.$$

(f) The fundamental group of a compact Riemann surface of genus g has $2g$ generators $u_1, v_1, ..., u_g, v_g$ and a single relation

$$u_1 v_1 u_1^{-1} v_1^{-1} \ldots u_g v_g u_g^{-1} v_g^{-1} = 1$$

Proposition: Let G be the group defined by the presentation (X, R). For any group H and map (of sets) $X \to H$ sending each element of R to 1 (in an obvious sense), there exists a unique homomorphism $G \to H$ making the following diagram commute:

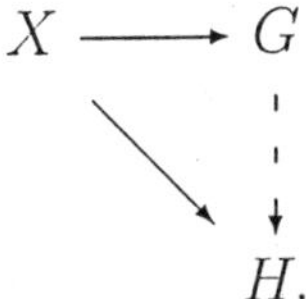

Proof: Let be a map $X \to H$. From the universal property of free groups (2.3), we know that σ extends to a homomorphism $FX \to H$, which we again denote. Let ιR be the image of R in FX. By assumption ιR Ker(α). Hence factors through $FX/N = G$. This proves the existence, and the uniqueness follows from the fact that we know the map on a set of generators for X.

Example: Let $G = \langle a, b | a^n, b^2, baba\rangle$. We prove that G is isomorphic to D_n. Because the elements $\sigma, \tau \in D_n$ satisfy these relations, the map

$$\{a, b\} \to D_n,\ a \mapsto \sigma\ b \mapsto \tau$$

extends uniquely to a homomorphism G $\to D_n$. This homomorphism is surjective because and τ generate D_n. The relations $a^n = 1$, $b^2 = 1$, $ba = a^{n-1}b$ imply that each element of G is represented by one of the following elements, 1, . . . , a^{n-1}, b, ab, . . . , $a^{n-1}b$, and so $(G : 1) \leq 2n = (D_n : 1)$. Therefore the homomorphism is bijective (and these symbols represent distinct elements of G).

Finitely Presented Groups

A group is said to be finitely presented if it admits a presentation (X, R) with both X and R finite.

Example: Consider a finite group G. Let $X = G$, and let R be the set of words

$$\{abc^{-1} \mid ab = c \text{ in } G\}.$$

We claim that (X,R) is a presentation of G, and so G is finitely presented. Let $G' = \langle X|R\rangle$. The map $FX \to G$, $a \to a$, sends each element of R to 1, and therefore defines a homomorphism $G' \to G$, which is obviously surjective. But clearly every element of G' is represented by an element of X, and so the homomorphism is also injective. Although it is easy to define a group by a finite presentation, calculating the properties of the group can be very difficult — note that we are defining the group, which may be quite small, as the quotient of a huge free group by a huge subgroup.

Let G be the group defined by a finite presentation (X, R). The word problem for G asks whether there is an algorithm (decision procedure) for deciding whether a word on X' represents 1 in G. Unfortunately, the answer is negative: Novikov and Boone showed that there exist finitely presented groups G for which there is no such algorithm. Of course, there do exist other groups for which there is an algorithm. The same ideas lead to the following result: there does not exist an algorithm that will determine for an arbitrary finite presentation whether or not the corresponding group is trivial, finite, abelian, solvable, nilpotent, simple, torsion, torsion-free, free, or has a solvable word problem.

The Burnside Problem

A group is said to have exponent m if $g^m = 1$ for all $g \in G$. It is easy to write down examples of infinite groups generated by a finite number of elements of finite order, but does there exist an infinite finitely-generated group with a finite exponent?.

There are some quite innocuous looking finite presentations that are known to define quite small groups, but for which this is very difficult to prove. The standard approach to these questions is to use the Todd-Coxeter algorithm. In the remainder of this course, including the exercises, we will develop various methods for recognizing groups from their presentations.

Isomorphism Theorems: Extensions

Theorems Concerning Homomorphisms

The next three theorems (or special cases of them) are often called the first, second, and third isomorphism theorems respectively.

Factorization of Homomorphisms

Recall that the image of a map : $S \to T$ is $\alpha(S) = \{\alpha(s) | s \varepsilon S\}$

Theorem 1. (Fundamental Theorem of group Homomorphisms) For any homomorphism : $G \to G'$ of groups, the kernel N of is a normal subgroup of G, the image I of is α subgroup of G', and factors in a natural way into the composite of a surjection, an isomorphism, and an injection:

$$\begin{array}{ccc} G & \xrightarrow{\alpha} & G' \\ \Big\downarrow \text{onto} & & \Big\uparrow \text{inj.} \\ G/N & \xrightarrow{\cong} & I \end{array}$$

Proof: We have seen that the kernel is a normal subgroup of G. If $b = \alpha(\mathrm{a})$ and $b' = \alpha(a')$, then $bb' = \alpha\ (aa')$ and $b^{-1} = \alpha(a^{-1})$, and so $I =_{df} \alpha\ (G)$ is a subgroup of G'. For $n\ \varepsilon\ N$, $\alpha(gn) = \alpha(g)$ $\alpha(n) = \alpha(g)$, and so α is constant on each left coset gN of N in G. It therefore defines a map

$$\overline{\alpha}\, G/N \to I,\ \overline{\alpha}\,(gN) = \alpha(g)$$

Then $\overline{\alpha}$ is a homomorphism because

$$\overline{\alpha}((gN)\cdot(g'N)) = \overline{\alpha}(gg'N) = \overline{\alpha}(gg') = \overline{\alpha}(g)\overline{\alpha}(g')$$

and it is certainly surjective. If $\overline{\alpha}(gN) = 1$, then $g \in \text{Ker}(\overline{\alpha}) = N$, and so $\overline{\alpha}$ has trivial kernel. This implies that it is injective.

The Isomorphism Theorem

Theorem 2: (Isomorphism Theorem). Let H be a subgroup of G and N a normal subgroup of G. Then HN is a subgroup of G, $H \cap N$ is a normal subgroup of H, and the map

$$h(H \cap N) \to hN : H/H \cap N \to HN/N$$

is an isomorphism.

Proof: We have already seen that HN is a subgroup. Consider the map

$$H \to G/N,\ h \to hN.$$

This is a homomorphism, and its kernel is $H \cap N$, which is therefore normal in H. According to Theorem 1, it induces an isomorphism $H/H \cap N \to I$ where I is its image. But I is the set of cosets of the form hN with $h \in H$, i.e., $I = HN/N$.

The Correspondence Theorem

The next theorem shows that if $\overline{G}$ is a quotient group of G, then the lattice of subgroups in $\overline{G}$ captures the structure of the lattice of subgroups of G lying over the kernel of $G \to \overline{G}$.

Theorem 3 (Correspondence Theorem). Let $\alpha\ G \to \overline{G}$ be a surjective homomorphism, and let $N = \text{Ker}(\alpha)$. Then there is a one-to-one correspondence

$$\{\text{subgroups of } G \text{ containing } N\} \xleftrightarrow{1:1} \{\text{subgroups of } \overline{G}\}$$

under which a subgroup H of G containing N corresponds to $\overline{H} = \overline{G}$ corresponds to $H = \alpha^{-1}(\overline{H})$. Moreover, if $H \leftrightarrow \overline{H}$ and $H' \leftrightarrow H'$, then

(a) $\overline{H} \subset H' \Leftrightarrow H \subset H'$, in which case $(\overline{H}' : \overline{H}) = (H' : H)$;

(b) $\overline{H}$ is normal in $\overline{G}$ if and only if H is normal in G, in which case, α induces an isomorphism

$$G/H \xrightarrow{\cong} \overline{G}/\overline{H}.$$

Proof: For any subgroup $\overline{H}$ of $\overline{G}$, $\alpha^{-1}(\overline{H})$ is a subgroup of G containing N, and for any subgroup H of G, $\alpha(H)$ is a subgroup of G. One verifies easily that $\alpha^{-1}\alpha(H) = H$ if and only if $H \supset N$, and that $\alpha\alpha^{-1}(\overline{H}) = \overline{H}$. Therefore, the two operations give the required bijection. The remaining statements are easily verified.

Corollary 4: Let N be a normal subgroup of G; then there is a one-to-one correspondence between the set of subgroups of G containing N and the set of subgroups of G/N, $H \leftrightarrow H/N$. Moreover H is normal in G if and only if H/N is normal in G/N, in which case the homomorphism $g \mapsto gN : G \to G/N$ induces an isomorphism

$$G/H \xrightarrow{\cong} (G/N)/(H/N).$$

Proof: Special case of the theorem in which α is taken to be $g \mapsto gN : G \to G/N$.

Direct Products

The next two propositions give criteria for a group to be a direct product of two subgroups.

Proposition 5: Consider subgroups H_l and H_2 of a group G. The map

$$(h_1, h_2) \mapsto h_1h_2 : H_1 \times H_2 \to G$$

is an isomorphism of groups if and only if

(a) $G = H_1H_2$,

(b) $H_1 \cap H_2 = \{1\}$, and

(c) every element of H_1 commutes with every element of H_2.

Proof: The conditions are obviously necessary (if $g \in H_1 \cap H_2$, then $(g, g^{-1}) \to 1$, and so $(g, g^{-1}) = (1, 1)$). Conversely, (c) implies that the map $(h_1, h_2) \to h_1h_2$ is a homomorphism, and (b) implies that it is injective:

$$h_1h_2 = 1 \Rightarrow h_1 = h_2^{-1} \in H_1 \cap H_2 = \{1\}.$$

Finally, (a) implies that it is surjective.

Proposition 6: Consider subgroups H_1 and H_2 of a group G. The map

$$(h_1, h_2) \mapsto h_1h_2 : H_1 \times H_2 \to G$$

is an isomorphism of groups if and only if

(a) $H_1H_2 = G$,

(b) $H_1 \cap H_2 = \{1\}$, and

(c) H_1 and H_2 are both normal in G.

Proof: Again, the conditions are obviously necessary. In order to show that they are sufficient, we check that they imply the conditions of the previous proposition. For this we only have to show that each element h_1 of H_1 commutes with each element h_2 of H_2. But the commutator $[h_1, h_2] = h_1h_2h^{-1}_1, h_2^{-1} = (h_1h_2h^{-1}_1) \cdot h_2^{-1}$ is in H_2 because H_2 is normal, and it's in H_1 because H_1 is normal, and so (b) implies that it is 1. Hence $h_1h_2 = h_2h_1$.

Proposition 7: Consider subgroups $H_1, H_2, \ldots, H_k$ of a group G. The map

$$(h_1, h_2, \ldots, h_k) \mapsto h_1h_2 \cdots h_k : H_1 \times H_2 \times \cdots \times H_k \to G$$

is an isomorphism of groups if (and only if)

(a) $G = H_1H_2 \cdots H_k$,

(b) for each j, $H_j \cap (H_1 \cdots H_{j-1}H_{j+1} \cdots H_k) = \{1\}$, and

(c) each of $H_1, H_2, \ldots, H_k$ is normal in G,

Proof: For $k = 2$, this is becomes the preceding proposition. We proceed by induction on k. The conditions (a,b,c) hold for the subgroups $H_1, \ldots, H_{k-1}$ of $H_1 \cdots H_{k-1}$, and so we may assume that

$$(h_1, h_2, \ldots, h_{k-1}) \mapsto h_1h_2 \cdots h_{k-1} :$$
$$H_1 \times H_2 \times \cdots \times H_{k-1} \to H_1H_2 \cdots H_{k-1}$$

is an isomorphism. An induction argument using shows that $H_1 \cdots H_{k-1}$ is normal in G, and so the pair $H_1 \cdots H_{k-1}$, H_k satisfies the hypotheses of. Hence

$$(h, h_k) \mapsto hh_k : (H_1 \cdots H_{k-1}) \times H_k \to G$$

is an isomorphism. These isomorphisms can be combined to give the required isomorphism:

$$H_1 \times \cdots \times H_{k-1} \times H_k \xrightarrow{(h_1,\ldots,h_k)\mapsto(h_1\ldots h_{k-1},h_k)}$$
$$H_1 \cdots H_{k-1} \times H_k \xrightarrow{(h,h_k)\mapsto hh_k} G.$$

Remark: 8. When

$$(h_1, h_2, \ldots, h_k) \mapsto h_1h_2 \cdots h_k : H_1 \times H_2 \times \cdots \times H_k \to G$$

is an isomorphism we say that G is the direct product of its subgroups H_i. In more down to-earth terms, this means: each element g of G can be written uniquely in the form $g = h_1h_2 \cdots h_k$, $h_i \in H_i$; if $g = h_1h_2 \cdots h_k$ and $g' = h'_1h'_2 \cdots h'_k$, then

$$gg' = (h_1h'_1)(h_2h'_2) \cdots (h_kh'k).$$

Automorphisms of Groups

Let G be a group. An isomorphism $G \to G$ is called an automorphism of G. The set Aut(G) of such automorphisms becomes a group under composition: the composite of two automorphisms is again an automorphism; composition of maps is always associative; the identity map $g \alpha g$ is an identity element; an automorphism is a bijection, and therefore has an inverse, which is again an automorphism.

For $g \in G$, the map i_g "conjugation by g",

$$x \mapsto gxg^{-1} : G \to G$$

is an automorphism: it is a homomorphism because

$$g(xy)g^{-1} = (gxg^{-1})(gyg^{-1}), \text{ i.e., } i_g(xy) = i_g(x)i_g(y),$$

and it is bijective because $i_{g^{-1}}$ is an inverse. An automorphism of this form is called an inner automorphism, and the remaining automorphisms are said to be outer.

Note that

$$(gh)x(gh)^{-1} = g(hxh^{-1})g^{-1}, \text{ i.e., } i_{gh}(x) = (i_g \circ i_h)(x),$$

and so the map $g \mapsto i_g : G \to \text{Aut}(G)$ is a homomorphism. Its image is written Inn(G). Its kernel is the centre of G,

$$Z(G) = \{g \in G \mid gx = xg \text{ all } x \in G\},$$

and so we obtain from (1) an isomorphism

$$G/Z(G) \to \text{Inn}(G).$$

In fact, Inn(G) is a normal subgroup of Aut(G): for $g \in G$ and $\alpha \in$ Aut(G),

$$(\alpha o i_g o \alpha^{-1})\ (x) = \alpha(g \cdot \alpha^{-1}\ (x) \cdot g^{-1}) = \alpha(g) \cdot x \cdot \alpha(g)^{-1} = i_\alpha(g)(x)$$

A group G is said to be complete if the map $g \mapsto i_g : G \to \text{Aut}(G)$ is an isomorphism. Note that this is equivalent to the condition:

(a) the centre $Z(G)$ of G is trivial, and

(b) every automorphism of G is inner.

Example 9.

(a) For $n \neq 2, 6$, S_n is complete. The group S_2 is commutative and hence fails (a); $\text{Aut}(S_6)/\text{Inn}(S_6) \approx C_2$, and hence S_6 fails (b).

(b) Let $G = F^n_p$. The automorphisms of G as an abelian group are just the automorphisms of G as a vector space over F_p; thus $\text{Aut}(G) = GL_n(F_p)$. Because G is commutative, all nontrivial automorphisms of G are outer.

(c) As a particular case of (b), we see that

$$\text{Aut}(C_2 \times C_2) = GL_2(F_2).$$

But $GL_2\ (F_2) \approx S_3$, and so the nonisomorphic groups $C_2 \times C_2$ and S_3 have isomorphic automorphism groups.

(d) Let G be a cyclic group of order n, say $G = \langle g \rangle$. An automorphism of G must send g to another generator of G. Let m be an integer ≥ 1. The smallest multiple of m divisible by n is $m \cdot \dfrac{n}{ged(m,n)}$. Therefore, g^m has order $\dfrac{n}{ged(m,n)}$ and so the generators of G are the elements g^m with $gcd\ (m,\ n) = 1$. Thus $\alpha\ (g) = g^m$ for some m relatively prime to n, and in fact the map $\alpha \mapsto m$ defines an isomorphism

$$\text{Aut}(C_n) \to (Z/nZ)^\times$$

where

$$(Z/nZ)\times = \{\text{units in the ring } Z/nZ\} = \{m + nZ \mid gcd(m, n) = 1\}.$$

This isomorphism is independent of the choice of a generator g for G; in fact, if $\alpha(g) = g^m$, then for any other element $g' = g^i$ of G,

$$\alpha(g') = \alpha(g^i) = \alpha(g)^i\ g^{mi} = (g^i)^m = (g')^m.$$

(e) Since the centre of the quaternion group Q is $\langle\alpha^2\rangle$, we have that

$$\text{Inn}(Q) \cong Q/\langle\alpha^2\rangle\ C_2 \times C_2.$$

In fact, $\text{Aut}(Q) \approx S_4$.

(f) If G is a simple noncommutative group, then Aut(G) is complete.

Remark 10: It will be useful to have a description of $(Z/nZ)^\times = \text{Aut}(C_n)$. If $n = p^{r1}{}_1 \cdots p^{rs}{}_1$ is the factorization of n into powers of distinct primes, then the Chinese Remainder Theorem gives us an isomorphism

$$Z/nZ \cong Z/p_1{}^{r1}Z \times ... \times Z/p_S{}^{rs}\, Z,\ m \bmod n \mapsto (m \bmod p_l{}^{rl}..., m \bmod p_S{}^{rs}),$$

which induces an isomorphism

$$(Z/nZ)^\times \cong (Z/p_1{}^{r1}Z)^\times \times \cdots \times (Z/p_S{}^{rs}Z)^\times.$$

Hence we need only consider the case $n = p^r$, p prime.

Suppose first that p is odd. The set $\{0, 1, \ldots, p^{r-l}\}$ is a complete set of representatives for Z/p^rZ, and $1/p$ of these elements are divisible by p. Hence $(Z/p^rZ)^\times$ has order $p^r - p^r/p = p^{r-1}(p-1)$. Because p^{-1} and p^r are relatively prime, we know from (3d) that $(Z/p^rZ)^\times$ is isomorphic to the direct product of a group A of order p^{-1} and a group B of order p^{r-1}. The map

$$(Z/p^rZ)^\times \mapsto (Z/pZ)^\times = F^\times_p ,$$

induces an isomorphism $A \to F^\times_p$, and $F^\times_p$, being a finite subgroup of the multiplicative group of a field, is cyclic (*FT*, Exercise 3). Thus $(Z/p^rZ)^\times \supset A = \langle\xi\rangle$ for some element ζ of order p-1. Using the binomial theorem, one finds that $1+p$ has order p^{r-1} in $(Z/p^rZ)^\times$, and therefore generates B. Thus $(Z/p^rZ)^\times$ is cyclic, with generator $\zeta\,(1 + p)$, and every element can be written uniquely in the form

$$\zeta \cdot (1 + p)^j , 0 \leq i < p - 1, 0 \leq j < p^{r-1}.$$

On the other hand,

$$(Z/8Z)^\times = \{ \bar{1},\bar{3},\bar{5},\bar{7} \} = \bar{3},\bar{5} \approx C_2 \times C_2$$

is not cyclic. The situation can be summarized by:

$$(Z/p^rZ)^\times \approx \begin{cases} C_{(p-1)pr-1} & p \text{ odd}, \\ C^2 & p^r = 22 \\ C_2 \times C_{2r-2} & p = 2, r > 2. \end{cases}$$

Definition: A characteristic subgroup of a group G is a subgroup H such that $\alpha\,(H) = H$ for all automorphisms α of G.

The same argument as in shows that it suffices to check that $\alpha\,(H) \subset H$ for all $\alpha \in \text{Aut}(G)$.

Contrast: A subgroup H of G is normal if it is stable under all inner automorphisms of G; it is characteristic if it stable under all automorphisms. In particular, a characteristic subgroup is normal.

Remark: (a) Consider a group G and a normal subgroup H. An inner automorphism of G restricts to an automorphism of H, which may be outer. Thus a normal subgroup of H need not be a normal subgroup of G. However, a characteristic subgroup of

H will be a normal subgroup of G. Also a characteristic subgroup of a characteristic subgroup is a characteristic subgroup.

(b) The centre $Z(G)$ of G is a characteristic subgroup, because

$$zg = gz \text{ all } g \in G \Rightarrow \alpha(z)\,\alpha(g) = \alpha(g)\alpha(z) \text{ all } g \in G,$$

and as g runs over G, $\alpha(g)$ also runs over G. Expect subgroups with a general grouptheoretic definition to be characteristic.

(c) If H is the only subgroup of G of order m, then it must be characteristic, because α (H) is again a subgroup of G of order m.

(d) Every subgroup of a commutative group is normal but not necessarily characteristic. For example, a subspace of dimension 1 in $G = F_p^2$ will not be stable under $GL_2(F_p)$ and hence is not a characteristic subgroup.

Semidirect Products

Let N be a normal subgroup of G. Each element g of G defines an automorphism of N, $n \mapsto gng^{-1}$, and so we have a homomorphism

$$\theta\ G \to \text{Aut}(N).$$

If there exists a subgroup Q of G such that $G \to G/N$ maps Q isomorphically onto G/N, then I claim that we can reconstruct G from the triple $(N, Q, \theta|Q)$. Indeed, any $g \in G$ can be written in a unique fashion

$$g = nq,\ n \in N,\ q \in Q$$

q is the unique element of Q representing g in G/N, and $n = gq^{-1}$. Thus, we have one-to-one correspondence (of sets)

$$G \xleftrightarrow{1-1} N \times Q.$$

If $g = nq$ and $g' = n'q'$, then

$$gg' = nqn'q' = n(qn'q^{-1})qq' = n \cdot \theta(q)(n') \cdot qq'.$$

Definition: A group G is said to be a semidirect product of the subgroups N and Q, written $N \times Q$, if N is normal and $G \to G/N$ induces an isomorphism $Q \xrightarrow{\approx} G/N$. Equivalent condition: N and Q are subgroups of G such that

$$\text{(i) } N \triangleleft G; \text{ (ii) } NQ = G; \text{ (iii) } N \cap Q = \{1\}.$$

Note that Q need not be a normal subgroup of G.

Example: (*a*) In D_n, let $C_n = \langle s \rangle$ and $C_2 = \langle \tau \rangle$; then

$$D_n = \langle \sigma \rangle \times \langle \tau \rangle = C_n \times C_2.$$

(b) The alternating subgroup A_n is a normal subgroup of S_n, and $Q = \{(12)\} \xrightarrow{\approx} S_n/A_n$. Therefore $S_n = A_n \times C_2$.

(c) The quaternion group can not be written as a semidirect product in any nontrivial fashion.

(d) A cyclic group of order p^2, p prime, is not a semidirect product.

(e) Let $G = GL_n(k)$, the group of invertible $n \times n$ matrices with coefficients in the field k. Let B be the subgroup of upper triangular matrices in G, T the subgroup of diagonal matrices in G, and U subgroup of upper triangular matrices with all their diagonal coefficients equal to 1. Thus, when $n = 2$,

$$B = \left\{\begin{pmatrix} * & * \\ 0 & * \end{pmatrix}\right\}, \quad T = \left\{\begin{pmatrix} * & 0 \\ 0 & * \end{pmatrix}\right\}, \quad U = \left\{\begin{pmatrix} 1 & * \\ 0 & 1 \end{pmatrix}\right\},$$

Then, U is a normal subgroup of B, $UT = B$, and $U \cap T = \{1\}$. Therefore,

$$B = U \times T.$$

Note that, when $n \geq 2$, the action of T on U is not trivial, and so B is not the direct product of T and U.

We have seen that, from a semidirect product $G = N \times Q$, we obtain a triple

$$(N, Q, \theta : Q \to \mathrm{Aut}(N)).$$

We now prove that every triple (N, Q, θ) consisting of two groups N and Q and a homomorphism $\theta : Q \to \mathrm{Aut}(N)$ arises from a semidirect product. As a set, let $G = N \times Q$, and define

$$(n, q)(n', q') = (n \cdot \theta(q)(n'), qq').$$

Proposition: The above composition law makes G into a group, in fact, the semidirect product of N and Q.

Proof: Write ${}^{q}n$ for $\theta(q)(n)$, so that the composition law becomes

$$(n, q)(n', q') = (n \cdot {}^{q}n', qq').$$

Then

$$((n, q), (n', q'))(n'', q'') =$$
$$(n \cdot {}^{q}n' \cdot {}^{qq'}n'', qq'q'') = (n, q)((n', q')(n'', q''))$$

and so the associative law holds.

Because $\theta(1) = 1$ and $\theta(q)(1) = 1$,

$$(1, 1)(n, q) = (n, q) = (n, q)(1, 1),$$

and so $(1, 1)$ is an identity element. Next

$$(n, q)({}^{q^{-1}}n, q^{-1}) = (1, 1) = ({}^{q^{-1}}n, q^{-1})(n, q),$$

and so $({}^{q^{-1}}n, q^{-1})$ is an inverse for (n, q). Thus G is a group, and it is easy to check that it satisfies the conditions.

Write $G = N \times Q$ for the above group.

Example: (a) Let θ be the (unique) nontrivial homomorphism

$$C_4 \to \mathrm{Aut}(C_3) \cong C_2,$$

namely, that which sends a generator of C_4 to the map $a \alpha a^2$. Then $G =_{df} C_3 \times_\theta C_4$ is a noncommutative group of order 12, not isomorphic to A_4. If we denote the generators of C_3 and C_4 by a and b, then a and b generate G, and have the defining relations

$$a^3 = 1,\ b^4 = 1,\ bab^{-1} = a^2.$$

(b) The bijection

$$(n, q) \mapsto (n, q) : N \times Q \to N \times_\theta Q$$

is an isomorphism of groups if and only if θ is the trivial homomorphism $Q \to \mathrm{Aut}(N)$, i.e., $(q)(n) = n$ for all $q \in Q$, $b \in N$.

(c) Both S_3 and C_6 are semidirect products of C_3 by C_2 — they correspond to the two homomorphisms $C_2 \to C_2 \cong \mathrm{Aut}(C_3)$.

(d) Let $N = \langle a, b\rangle$ be the product of two cyclic groups $\langle a\rangle$ and $\langle b\rangle$ of order p, and let $Q = \langle c\rangle$ be a cyclic group of order p. Define $\theta : Q \to \mathrm{Aut}(N)$ to be the homomorphism such that

$$\theta(c^i)(a) = ab^i,\ \theta(c^i)(b) = b.$$

[If we regard N as the additive group $N = F^2_p$ with a and b the standard basis elements, then $\theta(c^i)$ is the automorphism of N

defined by the matrix $\begin{pmatrix} 1 & 0 \\ i & 1 \end{pmatrix}$ 1.

The group $G =_{df} N \times_\theta Q$ is a group of order p^3, with generators a, b, c and defining relations

$$a^p = b^p = c^p = 1,\ ab = cac^{-1},\ [b, a] = 1 = [b, c].$$

Because $b \neq 1$, the group is not commutative. When p is odd, all elements except 1 have order p. When $p = 2$, $G \approx D_4$. Note that this shows that a group can have quite different representations as a semidirect product:

$$D_4 \overset{3.14a}{\approx} C_4 \times C_2 \approx (C_2 \times C_2) \times C_2.$$

(e) Let $N = \langle a \rangle$ be cyclic of order p^2, and let $Q = \langle b \rangle$ be cyclic of order p, where p is an odd prime. Then Aut $N \approx C_{p-1} \times C_p$, and the generator of C_p is α where $\alpha(a) = a^{1+p}$ (hence $\alpha^2(a) = a^{1+2p}, \ldots$). Define $Q \to \mathrm{Aut}N$ by $b \mapsto \alpha$. The group $G =_{\mathrm{df}} N \times Q$ has generators a, b and defining relations

$$a^{p2} = 1,\ b^p = 1,\ bab^{-1} = a^{1+p}.$$

It is a nonabelian group of order p^3, and possesses an element of order p^2.

For an odd prime p, the groups constructed in (d) and (e) are the only nonabelian groups of order p^3.

(f) Let α be an automorphism, possibly outer, of a group N. We can realize N as a normal subgroup of a group G in such a way that α becomes the restriction to N of an inner automorphism of G. To see this, let $\theta : C_\infty \to \mathrm{Aut}(N)$ be the homomorphism sending a generator α of C_∞ to $\alpha \in \mathrm{Aut}(N)$, and let $G = N \times C_\infty$. Then the element $g = (1, a)$ of G has the property that $g(n, 1)g^{-1} = \alpha\ (n), 1)$ for all $n \in N$. The semidirect product $N \times_\theta Q$ is determined by the triple

$$(N, Q,\ \theta : Q \to \mathrm{Aut}(N)).$$

It will be useful to have criteria for when two triples (N, Q, θ) and (N, Q, θ') determine isomorphic groups.

Leema: If θ and θ' are conjugate, i.e., there exists an $\alpha \in$ Aut(N) such that $\theta'(q) = \alpha\, o\theta(q) = o\, \alpha^{-1}$ for all $q \in Q$, then

$$N \times_{\theta} Q \approx N \times_{\theta} Q.$$

Proof: Consider the map

$$\gamma\colon N \times_{\theta} Q \to N \times_{\theta} Q,\ (n, q) \mapsto (\alpha(n), q).$$

Then

$$\begin{aligned} \gamma(n, q) \cdot \gamma(n', q') &= \alpha\,(n), q) \cdot (\alpha\,(n'), q') \\ &= \alpha\,(n) \cdot \theta'\,(q), \alpha\,(n')((qq') \\ &= (\alpha(n) \cdot (\alpha o \theta(q) o \alpha^{-1})(n'), qq') \\ &= (\alpha(n) \cdot \alpha(\theta(q)\,(n')), qq), \end{aligned}$$

and

$$\begin{aligned} \gamma\,(n, q)\,(n', q') &= \gamma\,(n.\ \theta(q)(n'), qq') \\ &= (\alpha(n).\ \alpha(\theta(q)(n')), qq'). \end{aligned}$$

Therefore γ is a homomorphism, with inverse $(n, q) \mapsto (\theta^{-1}(n), q)$, and so is an isomorphism.

Lemma: If $\theta = 0' \; o \; \theta$ with $\alpha \in$ Aut(Q), then

$$N \times Q \approx N \times_{\theta} Q.$$

Proof:. The map $(n, q) \mapsto (n, \alpha\,(q))$ is an isomorphism $N \times_{\theta} Q \to N \times_{\theta} o\; Q$.

Lemma: If Q is cyclic and the subgroup $\theta(Q)$ of Aut(N) is conjugate to $\theta(Q)$, then

$$N \times_{\theta} Q \approx N \times_{\theta'} Q.$$

Proof: Let a generate Q. Then there exists an i and an $\alpha \in$ Aut(N) such that

$$\theta'(a') = \alpha\ .\ \theta\,(a)\ .\ \alpha^{-1}$$

The map $(n, q) \mapsto (\alpha(n), q^{i})$ is an isomorphism $N \times_{\theta} Q \to N \times_{\theta'} Q$

Extensions of Groups

A sequence of groups and homomorphisms

$$1 \to N \xrightarrow{\iota} G \xrightarrow{\pi} Q \to 1$$

is exact if ι is injective, π is surjective, and $\mathrm{Ker}(\pi) = \mathrm{Im}(\iota)$. Thus $\iota(N)$ is a normal subgroup of G (isomorphic by ι to N) and $G/\iota(N) \xrightarrow{\approx} Q$. We often identify N with the subgroup $\iota(N)$ of G and Q with the quotient G/N.

An exact sequence as above is also referred to as an extension of Q by N. An extension is central if $\iota(N) \subset Z(G)$. For example,

$$1 \to N \to N \times_{\theta} Q \to Q \to 1$$

is an extension of N by Q, which is central if (and only if) θ is the trivial homomorphism.

Two extensions of Q by N are said to be isomorphic if there is a commutative diagram

$$\begin{array}{ccccccccc} 1 & \longrightarrow & N & \longrightarrow & G & \longrightarrow & Q & \longrightarrow & 1 \\ & & \| & & \downarrow \approx & & \| & & \\ 1 & \longrightarrow & N & \longrightarrow & G' & \longrightarrow & Q & \longrightarrow & 1. \end{array}$$

An extension

$$1 \to N \xrightarrow{\iota} G \xrightarrow{\pi} Q \to 1$$

is said to be split if it isomorphic to a semidirect product. Equivalent conditions:

(a) there exists a subgroup $Q' \subset G$ such that π induces an isomorphism $Q' \to Q$; or

(b) there exists a homomorphism s: $Q \to G$ such that $\pi \circ S = id$.

In general, an extension will not split. For example, the extensions

$$1 \to N \to Q \to Q/N \to 1$$

(N any subgroup of order 4 in the quaternion group Q) and

$$1 \to C_p \to C_{p2} \to C_p \to 1$$

do not split. We list two criteria for an extension to split.

Proposition: (Schur-Zsddenhaus Lemma). An extension of finite groups of relatively prime order is split.

Proposition: Let N be a normal subgroup of a group G. If N is complete, then G is the direct product of N with the centralizer of N in G,

$$C_G(N) \overset{df}{=} \{g \in G \mid gn = ng \text{ all } n \in N\}.$$

Proof: Let $Q = C_G(N)$. We shall check that N and Q satisfy the conditions of Proposition.

Observe first that, for any $g \in G$, $n \mapsto gng^{-1} : N \to N$ is an automorphism of N, and (because N is complete), it must be the inner automorphism defined by an element $\gamma = \gamma(g)$ of N; thus

$$gng^{-1} = \gamma n \gamma^{1} \text{ all } n \in N.$$

This equation shows that $\gamma^{1}g \in Q$, and hence $g = \gamma(\gamma^{1}g) \in NQ$. Since g was arbitrary, we have shown that $G = NQ$.

Next note that every element of $N \cap Q$ is in the centre of N, which (by the completeness assumption) is trivial; hence $N \cap Q = 1$.

Finally, for any element $g = ng \in G$,

$$gQg^{-1} = n(gQg^{-1})n^{-1} = nQn^{-1} = Q$$

(recall that every element of N commutes with every element of Q). Therefore Q is normal in G.

An extension

$$1 \to N \to G \to 1$$

gives rise to a homomorphism θ': $G \to \text{Aut}(N)$, namely,

$$\theta'(g)(n) = gng^{-1}$$

Let $\bar{q} \in G$ map to q in Q; then the image of $\theta'(\bar{q})$ in Aut (N)/ Inn(N) depends only on q; therefore we get a homomorphism

$$\theta: Q \to \text{Out}(N) \overset{df}{=} \text{Aut}(N)/\text{Inn}(N)$$

This map θ depends only on the isomorphism class of the ectension, and we write $\text{Ext}^1(G, N)_\theta$ for the set of isomorphism classes of extensions with a given θ. These sets have been extensively studied.

Groups Acting on Sets

General Definitions and Results

Definition: Let X be a set and let G be a group. A left action of G on X is a mapping $(g, x) \mapsto gx$: $G \times X \to X$ such that

(a) $1x = x$, for all $x \in X$;

(b) $(g_1g_2)x = g_1(g_2x)$, all $g_1, g_2 \in G$, $x \in X$.

A set together with a (left) action of G is called a (left) G-set.

The axioms imply that, for each $g \in G$, left translation by g,

$$gL : X \in X, x \mapsto gx,$$

has $(g^{-1})_L$ as an inverse, and therefore g_L is a bijection, i.e., $g_L \in$ Sym(X). Axiom (b) now says that

$$g \mapsto g_L : G \to \mathrm{Sym}(X)$$

is a homomorphism. Thus, from a left action of G on X, we obtain a homomorphism $G \to \mathrm{Sym}(X)$, and, conversely, every such homomorphism defines an action of G on X.

Example:(a) The symmetric group S_n acts on $\{1, 2, ..., n\}$. Every subgroup H of S_n acts on $\{1, 2, \ldots, n\}$.

(b) Every subgroup H of a group G acts on G by left translation,

$$H \times G \to G, (h, x) \mapsto hx.$$

(c) Let H be a subgroup of G. If C is a left coset of H in G, then so also is gC for any $g \in G$. In this way, we get an action of G on the set of left cosets:

$$G \times G/H \to G/H, (g, C) \mapsto gC.$$

(d) Every group G acts on itself by conjugation:

$$G \times G \to G, (g, x) \mapsto g_x =_{df} gxg^{-1}.$$

For any normal subgroup N, G acts on N and G/N by conjugation.

(e) For any group G, $\mathrm{Aut}(G)$ acts on G.

A right action $X \times G \to G$ is defined similarly. To turn a right action into a left action, set $g * x = xg^{-1}$. For example, there is a natural right action of G on the set of right cosets of a subgroup H in G, namely, $(C, g) \mapsto Cg$, which can be turned into a left action $(g, C) \mapsto Cg^{-1}$.

A morphism of G-sets (better G-map; G-equivariant map) is a map $\varphi\ X \to Y$ such that

$$\varphi(gx) = g\varphi(x), \text{ all } g \in G, x \in X.$$

An isomorphism of G-sets is a bijective G-map; its inverse is then also a G-map.

Orbits

Let G act on X. A subset $S \subset X$ is said to be stable under the action of G if

$$g \in G, x \in S \Rightarrow gx \in S.$$

The action of G on X then induces an action of G on S.

Write $x \sim G\ y$ if $y = gx$, some $g \in G$. This relation is reflexive because $x = 1x$, symmetric because

$$y = gx \Rightarrow x = g^{-1}y$$

(multiply by g^{-1} on the left and use the axioms), and transitive because

$$y = gx,\ z = g'y\ )\ z = g'(gx) = (g'g)x.$$

It is therefore an equivalence relation. The equivalence classes are called G-orbits. Thus the G-orbits partition X. Write $G\backslash X$ for the set of orbits.

By definition, the G-orbit containing x_0 is

$$Gx_0 = \{gx_0 \mid g \in G\}.$$

It is the smallest G-stable subset of X containing x_0.

Example: (a) Suppose G acts on X, and let $\alpha \in G$ be an element of order n. Then the orbits of $\langle a \rangle$ are the sets of the form

$$\{x', x', \ldots, \alpha^{n-1}x_0\}.$$

(These elements need not be distinct, and so the set may contain fewer than n elements.)

(b) The orbits for a subgroup H of G acting on G by left multiplication are the right cosets of H in G. We write $H\backslash G$ for the set of right cosets. Similarly, the orbits for H acting by right multiplication are the left cosets, and we write G/H for the set of left cosets. Note that the group law on G will not induce a group law on G/H unless H is normal.

(c) For a group G acting on itself by conjugation, the orbits are called conjugacy classes: for $x \in G$, the conjugacy class of x is the set

$$\{gxg^{-1} \mid g \in G\}$$

of conjugates of x. The conjugacy class of x' consists only of x' if and only if x' is in the centre of G. In linear algebra the conjugacy classes in $G = GL_n(k)$ are called similarity classes, and the theory of (rational) Jordan canonical forms provides a set of representatives for the conjugacy classes: two matrices are similar (conjugate) if and only if they have essentially the same Jordan canonical form. Note that a subset of X is stable if and only if it is a union of orbits. For example, a subgroup H of G is normal if and only if it is a union of conjugacy classes.

The group G is said to act transitively on X if there is only one orbit, i.e., for any two elements x and y of X, there exists a $g \in G$ such that $gx = y$. For example, S_n acts transitively on $\{1, 2, ...n\}$. For any subgroup H of a group G, G acts transitively on G/H. But G (almost) never acts transitively on G (or G/N or N) by conjugation.

The group G acts doubly transitively on X if for any two pairs (x, x'), (y, y') of elements of X with $x \neq x'$ and $y \neq y'$, there exists a (single) $g \in G$ such that $gx = y$ and $gx' = y'$. Define k-fold transitivity, $k \geq 3$, similarly.

Stabilizers

The stabilizer (or isotropy group) of an element $x \in X$ is

$$\text{Stab}(x) = \{g \in G \mid gx = x\}.$$

It is a subgroup, but it need not be a normal subgroup. In fact:

Lemma: If $y = gx$, then $\mathrm{Stab}(y) = g \cdot \mathrm{Stab}(x) \cdot g^{-1}$.

Proof. Certainly, if $g'x = x$, then

$$(gg'g^{-1})y = gg'x = gx = y.$$

Hence $\mathrm{Stab}(y) \supset g \cdot \mathrm{Stab}(x) \cdot g^{-1}$. Conversely, if $g'y = y$, then

$$(g^{-1}g'g)x = g^{-1}g'(y) = g^{-1}y = x,$$

and so $g^{-1}g'g \in \mathrm{Stab}(x)$, i.e., $g' \in g \cdot \mathrm{Stab}(x) \cdot g^{-1}$.

Clearly

$$\mathrm{Stab}(x) = \mathrm{Ker}(G \to \mathrm{Sym}(X)),$$

which is a normal subgroup of G. If $\bigcap \mathrm{Stab}(x) = \{1\}$, i.e., $G \to \mathrm{Sym}(X)$, then G is said to act effectively (or faithfully). It acts freely if $\mathrm{Stab}(x) = 1$ for all $x \in X$, i.e., if $gx = x \to g = 1$.

Exmple: (a) Let G act on G by conjugation. Then

$$\mathrm{Stab}(x) = \{g \in G \mid gx = xg\}.$$

This group is called the centralizer $C_G(x)$ of x in G. It consists of all elements of G that commute with, i.e., centralize, x. The intersection

$$C_G(x) = \{g \in G \mid gx = xg \ \forall x \in G\}$$

is a normal subgroup of G, called the centre $Z(G)$ of G. It consists of the elements of G that commute with every element of G.

(b) Let G act on G/H by left multiplication. Then Stab(H) = H, and the stabilizer of gH is gHg^{-1}.

For a subset S of X, we define the stabilizer of S to be

$$\text{Stab}(S) = \{g \in G \mid gS = S\}.$$

The same argument as in the previous proof of shows that

$$\text{Stab}(gS) = g \cdot \text{Stab}(S) \cdot g^{-1}.$$

Example: Let G act on G by conjugation, and let H be a subgroup of G. The stabilizer of H is called the normalizer *NG(H)* of H in G:

$$NG(H) = \{g \in G \mid gHg^{-1} = H\}.$$

Clearly *NG(H)* is the largest subgroup of G containing H as a normal subgroup.

The element $g \notin N_G(H)$ even though $gHg^{-1} \subset H$.

Transitive Actions

Proposition: Suppose G acts transitively on X, and let $x' \in X$; then

$$gH \mapsto gx_0 : G/\,\text{Stab}(x_0) \to X$$

is an isomorphism of G-sets.

Proof: It is well-defined because if $h, h' \in \text{Stab}(x_0)$, then $ghx_0 = gx_0 = gh'x_0$ for any $g \in G$. It is injective because

$$gx_0 = g'x_0 \Rightarrow g^{-1}g'x_0 = x_0 \Rightarrow g,\ g'$$

lie in the same left coset of Stab(x_0).

It is surjective because G acts transitively. Finally, it is obviously G-equivariant.

The isomorphism is not canonical: it depends on the choice of $x_0 \in X$. Thus to give a transitive action of G on a set X is not the same as to give a subgroup of G.

Corollary: Let G act on X, and let $O = Gx_0$ be the orbit containing x_0. Then the number of elements in O is

$$\#O = (G : \mathrm{Stab}(x_0)).$$

For example, the number of conjugates gHg^{-1} of a subgroup H of G is $(G: N_G(H))$.

Proof: The action of Gon O is transitive, and so $g \mapsto gx_0$ defines a bijection $G/\mathrm{Stab}(x_0) \to Gx_0$.

This equation is frequently useful for computing $\#O$.

Proposition: If G acts transitively on X, then, for any $x_0 \in X$,

$$\mathrm{Ker}(G \to \mathrm{Sym}(X))$$

is the largest normal subgroup contained in $\mathrm{Stab}(x_0)$.

Proof: Let $x_0 \in X$. Then

$$\mathrm{Ker}(G \to \mathrm{Sym}(X)) = \bigcap_{x \in X} \mathrm{Stab}(x) =$$

$$\bigcap_{g \in G} \mathrm{Stab}(gx_0) \overset{4.4}{=} \bigcap g\,\mathrm{Stab}(x_0) \cdot g^{-1}.$$

Hence, the proposition is a consequence of the following lemma.

Lemma. For any subgroup H of a group G, gHg^{-1} is the largest normal subgroup contained in H.

Proof: Note that $N_0 =_{\mathrm{df}} gHg^{-1}$, being an intersection of subgroups, is itself a subgroup. It is normal because

$$g_1 N_0 g_1^{-1} = \bigcap_{g \in G} (g_1 g) N_0 (g_1 g)^{-1} = N_0$$

for the second equality, we used that, as g runs over the elements of G, so also does g_1g. Thus N_0 is a normal subgroup of G contained in $1H1^{-1} = H$. If N is a second such group, then

$$N = gNg^{-1} \subset gHg^{-1}$$

for all $g \in G$, and so

$$N \subset \bigcap_{g \in G} gHg^{-1} = N_0.$$

The Class Equation

When X is finite, it is a disjoint union of a finite number of orbits:

$$X = \bigcup_{i=1}^{m} O_i \quad \text{(disjoint union)}.$$

Hence:

Proposition: The number of elements in X is

$$\#X = \sum_{i=1}^{m} \#O_i = \sum_{i=1}^{m} (G : \text{Stab}(x^i)), \; C_G(x))$$

When G acts on itself by conjugation, this formula becomes:

Proposition: (Class Equation).

$$(G : 1) = \Sigma(G : C_G(x))$$

(x runs over a set of representatives for the conjugacy classes), or

$$(G : 1) = (Z(G) : 1) + \Sigma(G : C_G(y))$$

(y runs over set of representatives for the conjugacy classes containing more than one element).

Theorem (Cauchy): If the prime p divides $(G : 1)$, then G contains an element of order p.

Proof: We use induction on $(G : 1)$. If for some y not in the centre of G, p does not divide $(G : C_G(y))$, then $p|C_G(y)$ and we can apply induction to find an element of order p in $C_G(y)$. Thus we may suppose that p divides all of the terms $(G : C_G(y))$ in the class equation (second form), and so also divides $Z(G)$. But $Z(G)$ is commutative, and it follows from the structure theorem of such groups that $Z(G)$ will contain an element of order p.

Corollary: Any group of order $2p$, p an odd prime, is cyclic or dihedral.

Proof: From Cauchy's theorem, we know that such a G contains elements τ and σ of orders 2 and p respectively. Let $H = \langle\sigma\rangle$. Then H is of index 2, and so is normal. Obviously $\tau \notin H$, and so $G = H\ [H \cup H\tau]$:

$$G = \{1, \sigma \ldots, \sigma^{p-1}, \tau, \sigma\tau, \ldots, \sigma^{p-1}\tau\}.$$

As H is normal, $\tau\sigma\tau^{-1} = \sigma^i$, some i. Because $\tau^2 = 1$, $\sigma = \tau^2\sigma\tau^{-2} = \tau(\tau\sigma\tau^{-1})\tau^{-1} = \sigma^{i2}$, and so $i^2 \equiv$ mod p. The only elements of F_p with square 1 are ± 1, and so $i \equiv 1$ or -1 mod p. In the first case, the group is commutative (any group generated by a set of commuting elements is obviously commutative); in the second $\tau\sigma\tau^{-1} = \sigma^{-1}$ and we have the dihedral group.

p-groups

Theorem:. A finite p-group $\neq 1$ has centre $\neq \{1\}$.

Proof: By assumption, $(G : 1)$ is a power of p, and it follows that $(G : C_G(y))$ is power of $p \neq p^0$) for all y in the class equation (second form). Since p divides every term in the class equation except (perhaps) $(Z(G) : 1)$, it must divide $(Z(G): 1)$ also.

Corollary: A group of order p^m has normal subgroups of order p^n for all $n \leq m$.

Proof: We use induction on m. The centre of G contains an element g of order p, and so $N = \langle g \rangle$ is a normal subgroup of G of order p. Now the induction hypothesis allows us to assume the result for G/N, and the correspondence theorem then gives it to us for G.

Proposition: A group of order p^2 is commutative, and hence is isomorphic to $C_p \times C_p$ or C^{p2}.

Proof: We know that the centre Z is nontrivial, and that G/Z therefore has order 1 or p. In either case it is cyclic, and the next result implies that G is commutative.

Lemma: Suppose G contains a subgroup H in its centre (hence H is normal) such that G/H is cyclic. Then G is commutative.

Proof: Let $a \in G$ be such that aH generates G/H, so that $G/H = \{(aH)^i \mid i \in Z\}$. Since $(aH)^i = a^iH$, we see that every element of G can be written $g = a^ih$ with $h \in H$, $i \in$ ©. Now

$$a^ih \cdot a^{i'}h' = a^ia^{i'}hh' \text{ because } H \subset Z(G)$$
$$= a^{i'}a^ih'h$$
$$= a^{i'}h' \cdot a^ih.$$

Remark: The above proof shows that if $H \subset Z(G)$ and G contains a set of representatives for G/H whose elements commute, then G is commutative.

It is now not difficult to show that any noncommutative group of order p^3 is isomorphic to exactly one of the groups constructed. Thus, up to isomorphism, there are exactly two noncommutative groups of order p^3.

Action on the Left Cosets

The action of G on the set of left cosets G/H of H in G is a very useful tool in the study of groups. We illustrate this with some examples.

Let $X = G/H$. Recall that, for any $g \in G$,

$$\mathrm{Stab}(gH) = g\,\mathrm{Stab}(H)g^{-1} = gHg^{-1}$$

and the kernel of

$$G \to \mathrm{Sym}(X)$$

is the largest normal subgroup gHg^{-1} of G contained in H.

Remark: (a) Let H be a subgroup of G not containing a normal subgroup of G other than 1. Then $G \to \mathrm{Sym}(G/H)$ is injective, and we have realized G as a subgroup of a symmetric group of order much smaller than $(G : 1)!$. For example, if G is simple, then the Sylow theorems imply that G has many proper subgroups $H \neq 1$ (unless G is cyclic), but (by definition) it has no such normal subgroup.

(b) If $(G : 1)$ does not divide $(G : H)!$, then

$$G \to \mathrm{Sym}(G/H)$$

cannot be injective, and we can conclude that H contains a normal subgroup $\neq 1$ of G. For example, if G has order 99, then it will have a subgroup N of order 11, and the subgroup must be normal. In fact, $G = N \times Q$.

Example: Corollary above shows that every group G of order 6 is either cyclic or dihedral. Here we present a slightly different argument. According to Cauchy's theorem, G must contain an element σ of order 3 and an element τ of order 2. Moreover $N =_{df} \langle\sigma\rangle$ must be normal because 6 does not divide 2! (or simply because it has index 2). Let $H = \langle t \rangle$.

Either (a) H is normal in G, or (b) H is not normal in G. In the first case, $\sigma\tau\sigma^{-1} = \tau$, i.e., $\sigma\tau = \tau\sigma$, and so shows that G is commutative, $G \approx C_2 \times C_3$. In the second case, $G \to \mathrm{Sym}(G/H)$ is injective, hence surjective, and so $G \approx S_3$.

5

The Molecular Symmetry Group

Each group has a specific set of irreducible representations with names such as A_1, B_g, E_0, and F_2. In the same way as we label a state according to its parity by determining the effect of E^* on the wave function of the state, we can label a state according to the irreducible representations of a symmetry group of the molecule in question by determining the effect of the elements of the group on the wave function. A symmetry group of a molecule is a group whose elements commute with the Hamiltonian of the molecule; using somewhat looser language the elements of a symmetry group of a molecule do not change the energy of the molecule. This does not seem to have anything to do with the customary use of the word 'symmetry' which is concerned with the geometrical shape of an object.

Different molecules have different symmetry groups, and also have different irreducible representations which are given in character tables. The irreducible representation labels are straightforward to determine and once we have labeled the energy levels in this way the molecule becomes much easier to understand. The elements of the CNPI group of a molecule all commute with the Hamiltonian of the isolated molecule in free space, and thus the CNPI group of a molecule is a symmetry group of that molecule. We could use the irreducible representations of the CNPI group of a molecule to label its energy levels, but this is not always a clever thing to do. It is better to use a subgroup of the CNPI group called the molecular symmetry group.

The definition of the molecular symmetry (MS) group is easier to appreciate after we have considered the question \what do we do with the symmetry labels"? In general terms we use them to label energy levels that can be distinguished in the experimental results that we are interpreting, and we donot need to do more. We can say that if symmetry group permits this then it is sufficiently a large symmetry group and it provides sufficient symmetry labeling.

The molecular symmetry group, in conjunction with the spatial three-dimensional pure rotation group *K*(spatial), allows this. To be more specific, we use the symmetry labels to determine nuclear spin statistical weights, and to identify all zero order levels that can and cannot interact as a result of considering (a) the effect of previously neglected terms in the complete Hamiltonian, or (b) the effect of applying an external perturbation such as an electric ormagnetic field. We also use the symmetry labels to decide which states of the molecule can be connected by transitions in the presence of electromagnetic radiation. We can use the molecular symmetry group and the group *K*(spatial) for these purposes.

Drawbacks of Using the CNPI Group

Let us consider the number of elements in the CNPI group. For a series of molecules the order of the CNPI group (twice the order of the CNP group) is as follows:

H_2	$2! \times 2 = 4;$	C_2H_6	$2! \times 6! \times 2 = 2880;$
H_2O	$2! \times 2 = 4;$	C_2H_5OH	$2! \times 6! \times 2 = 2880;$
BF_3	$3! \times 2 = 12;$	C_6H_6	$6! \times 6! \times 2 = 1036800;$

CH_3F	$3! \times 2 = 12$;	CH_3COCH_2 CH_2OH	$4! \times 8!\ 2! \times 2 = 3870720$;
CH_4	$4! \times 2 = 48$;	$C_6H_5CH_3$	$7! \times 8! \times 2 \approx 4 \times 10^8$:
C_2H_4	$2! \times 4! \times 2 = 96$;	$(C_6H_6)(H_2O)_2$	$6! \times 10! \times 2! \times 2 \approx 10^{10}$:
SF_6	$6! \times 2 = 1440$;	C_{60}	$60! \times 2 \approx 10^{82}$:

Clearly the order of the CNPI group can be very large, and it only depends on the chemical formula of the molecule; the structure of the molecule is unimportant. As well as being very large the CNPI group almost invariably produces a symmetry classification of the levels of a molecule in which there are systematic multiple labels.

The CNPI group of methane G_{48} has 10 irreducible representations and it provides a convenient example. Using G_{48} it turns out that the rotation-vibrational energy levels of a methane molecule would be labeled as being of one of the following five types: $A_1^+ \oplus A_1^- \oplus$, $A_2^+ \oplus A_2^+$, $E^+ \oplus E_{,,}$ $F_1^+ \oplus F_1^-$, or $F_2^+ \oplus F_2^-$ where the '+' and – 'parity labels given by the effect of the operation *E**. Such "double" labels indicate an accidental degeneracy. Such systematic degeneracies are not really accidental but we call them accidental here since they are not required by the symmetry of the CNPI group.

These degeneracies, which we will now call structural degeneracies, are caused by the presence of more than one version of the equilibrium structure, in a given electronic state of a molecule. We will first explain what we mean by 'versions of the equilibrium structure', and then show how structural degeneracy arises.

The idea of versions of an equilibrium structure can be explained using the methane molecule as an example. In Fig. 1a a methane molecule is shown with the nuclei in the equilibrium configuration for the ground electronic state. If we deform the molecule through the planar conguration (1b) we can invert the molecule to obtain the form shown in Fig. 1c.

The configurations in Figs. 1a and 1c are the nuclear configurations at two identically shaped deep minima in the potential energy surface V_N of the ground electronic state, and the conguration shown in Fig. 1b represents a saddle point in V_N between the minima. Figs. 1a and 1c show the two possible versions of the methane molecule in its equilibrium conguration.

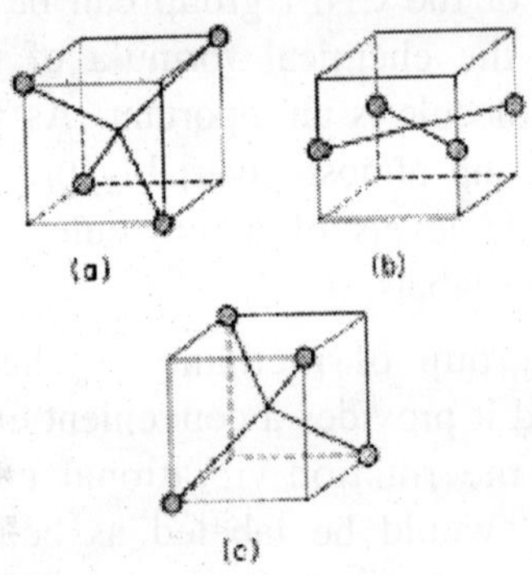

Fig. 1. A methane molecule in two different symmetrically equivalent equilibrium structures, (a) and (c), which can be interconverted by a deformation through the planar configuration shown in (b).

Different versions of the equilibrium structure arise not only in molecules, such as methane, in which they can be interchanged by inversion, but also rather commonly in molecules such as ethane (H_3CCH_3) in which they can be interchanged by torsion. There are other, less common, ways for them to arise, as we shall see when we consider the ethylene molecule. Introducing some other terminology that we will use, we can say that when a molecule contorts from one version to another it undergoes a degenerate rearrangement.

To distinguish between the different versions of the equilibrium structure of a molecule it is necessary to label the nuclei. Having labeled the nuclei of a molecule in its equilibrium structure, we determine the number of versions of the equilibrium structure by finding out how many distinct forms can be obtained by permuting the labels on identical nuclei with and without inverting the molecule. Versions are such that to interconvert them one cannot merely rotate the molecule in space but one must deform the molecule across a potential barrier. For the methane molecule there are only two versions of the equilibrium structure as we can see once we have labeled the protons (see Fig. 2). We call the versions in Figs. 2a and 2b the *A* (for anticlockwise) version and the C (for clockwise) version, respectively, since in the A version 1 → 2 → 3 is anticlockwise (clockwise) looking in the C → H_4 direction.

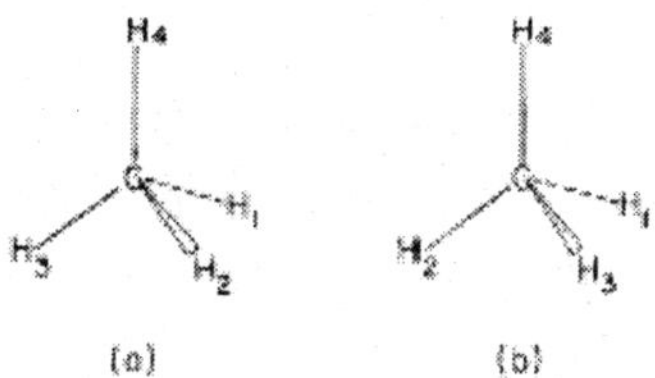

Fig. 2. The two versions of the equilibrium structure of a methane molecule. They are distinguishable because we have numbered the protons. The version in (a) is called the anticlockwise (A) version in the text because 1 → 2 → 3 is anticlock- wise when viewed in the C → H_4 direction, and (b) is the clockwise (C) version.

Labeling the protons one to four and the carbon nuclei five and six in an ethylene molecule, we see that in its electronic ground state there are altogether 12 distinct numbered versions of the equilibrium structure. To interchange pairs of these versions we either have to twist one CH_2 group relative to the other, or we have to break bonds while reforming new ones.

To understand structural degeneracy, and how it arises when there are versions of the equilibrium structure, we need only have a qualitative appreciation of the solution of the vibrational Schrodinger equation. For the methane molecule we can choose either structure *A* or *C* in Fig. 2 as the equilibrium conguration for the purpose of dening the vibrational displacements.

Depending upon which structure (*A* or *C*) we choose we obtain vibrational wavefunctions $\Phi^{(n)}_A$ and energies $E^{(n)}_A$, or $\Phi^{(n)}_C$ and $E^{(n)}_C$, where $n = 1; 2; 3; : : :$;for the successive eigenstates. If, as is true in methane, the barrier in V_N between the minima *A* and *C* is very high then the wavefunctions $\Phi^{(n)}_A$ are localized in the A minimum and the wavefunctions $\Phi^{(n)}_C$ in the *C* minimum, with no effective penetration of either into the other minimum. In other words the molecular vibrations just occur in the region around each minimum.

Since the two minima in V_N for methane have the same shape (they are symmetrically equivalent) the energies $E^{(n)}_A$ and $E^{(n)}_C$ will be identical (and identical to the observed vibrational energies $E^{(n)}$). Thus each observed vibrational energy level will be doubly degenerate, corresponding to the energies $E^{(n)}_A$ and $E^{(n)}_C$, and this is structural degeneracy. If there are n versions of the global minimum (at the equilibrium structure) in the potential function V_N for a molecule, with no effective penetration (tunnelling) of the barriers between these minima by the local wavefunctions, then each level will be *n*-fold structurally degenerate. When tunneling occurs the degeneracy is split, and this happens if the barrier is not high relative to the vibrational energy.

In actual fact there will always be some tunneling, since potential energy barriers are not infinitely high, but often the experimental resolution is not high enough to detect the splitting. This is the case at the moment for methane. Structural degeneracy is not required by the symmetry of the CNPI group since the symmetry group is the same regardless of the height of the barrier to tunneling, and the possibility of the splitting must therefore be allowed for by the symmetry labels obtained using the CNPI group. A molecule for which no observable tunneling between minima in V_N occurs is said to be rigid, and one for which observable tunneling occurs is said to be non rigid. It is important to realize that this definition of a rigid molecule still allows non zero vibrational amplitudes and it allows centrifugal distortion.

No experimental splittings, torsional or otherwise, resulting from tunneling through the barriers that separate the 12 versions of ethylene have been observed; the barriers in the ground electronic state potential function that separate these 12 versions are for all practical purposes insuperable and the molecule is, therefore, a rigid molecule. To determine the vibrational energy level pattern of ethylene we would only need to consider one numbered version and the shape of the one deep minimum containing that version in the potential energy surface V_N. Each of the twelve versions would have identical vibrational energy levels and these would match the observed vibrational energy levels of the molecule. Thus each of the vibrational energy levels of ethylene in its electronic ground state has a twelve-fold structural degeneracy.

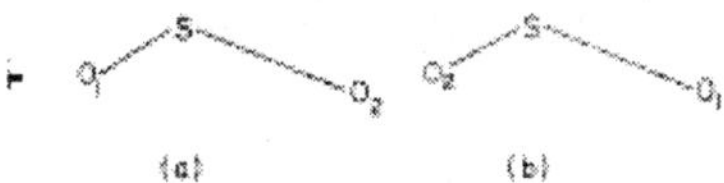

Fig. 3. The two versions of an unsymmetrical bent SO_2 molecule.

Structural degeneracy is present in nearly all molecules that contain identical nuclei. The lower the structural symmetry of a molecule the more versions of the equilibrium structure there will be, and the structural degeneracy increases at a dramatic rate with the size of the molecule. Simple symmetrical molecules such as SO_2 or BF_3 in their ground electronic states have no structural degeneracy since for each there is only one distinct numbered version.

However, it is possible that SO_2 has unequal bond lengths at equilibrium in an excited electronic state; if true this means that in this excited electronic state there would be two distinct numbered versions of the molecule at equilibrium, as shown in Fig. 3, and each level would be structurally doubly degenerate if there was no tunneling between the versions. If there were an excited electronic state of BF_3 in which the molecule were planar with three unequal bond lengths at equilibrium, with no observable tunneling splittings, then each level would have a structural degeneracy of six. In methane the vibrational levels of a nonplanar electronic state with four unequal equilibrium bond lengths and no tunneling splittings, would have a structural degeneracy of 48.

We see that, although mathematically correct, the use of the CNPI group has the drawbacks of often being incredibly large and often producing asymmetry labeling in which accidental degeneracies occur in a systematic fashion. We only need the chemical formula of a molecule in order to set up the CNPI group, and although we often know, or can guess, the equilibrium structure of a molecule and the tunneling possibilities, this information is not used in setting up the group. This information is used in setting up the molecular symmetry group, and the molecular symmetry group does not suffer from these drawbacks.

The Definition of the Molecular Symmetry (MS) Group

To appreciate the definition of the molecular symmetry (MS) group let us consider the problem of calculating the vibrational energies of the methyluoride molecule. The potential energy surface V_N for the methyluoride molecule in its ground electronic state has two deep minima centered at the anticlockwise (A) and clockwise (C) labeled versions shown in Figs. (4a) and (4b), respectively. To determine the vibrational energy levels a mathematician might insist that we should determine the eigen values of the vibrational Hamiltonian using the complete double minimum potential energy surface. However, experience tells us that there will be no observable splittings as a result of tunneling between the A and C versions for the vibrational levels that we study. Thus for all practical purposes we can determine the vibrational energy levels of the methyluoride molecule by solving the vibrational equation for the energy levels in one minimum, appropriate for the A version, say, with the complete neglect of the potential energy surface in the region of the other minimum. Vibrational interactions or external perturbations will only connect one A version level with another A version level in the absence of inversion tunneling. The C version will provide a duplicate set of energy levels and interactions so that the methyluoride molecule has two-fold structural degeneracy.

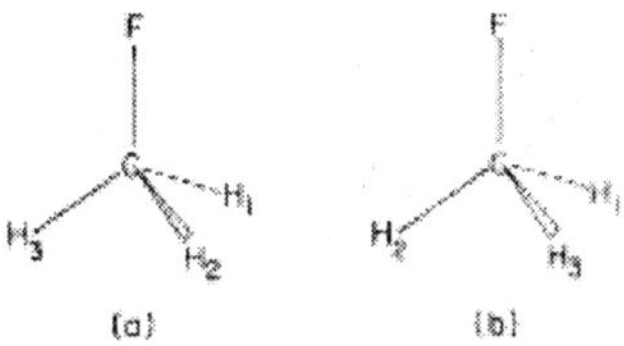

Fig. 4. The two versions of a methyluoride molecule at equilibrium where (a) is the anticlockwise (A) version and (b) is the clockwise (C) version.

We can completely understand the rotation-vibration energy levels and all possible interactions within the ground electronic state of methyluoride (a part-from inversion tunneling) by considering only one version of the molecule. Hence we need only make a symmetry labeling of the levels of one version to obtain a sufficient symmetry labeling of the levels of the molecule. To do this we just need the CNPI group of one version of the molecule, i.e., the group

$$\{E; (123); (132); (12)^*; (23)^*; (13)^* \qquad (1)$$

this group, called $C_{3v}(M)$, was introduced in Eq. (13) in the answer to Problem 2. The $C_{3v}(M)$ group in Eq. (1) is the molecular symmetry group of methyluoride when inversion tunneling splittings are not observed. Elements in the CNPI group of methyluoride that interconvert the A and C versions, such as (12) or (123), are not present in the MS group and constitute what are called unfeasible elements of the CNPI group of methyluoride; we could equally well call them useless elements of the CNPI group.

The use of the word 'unfeasible' for a permutation or permutation-inversion caused some controversy in the early days after Longuet-Higgins' paper; it seemed to suggest a 'pathway' for such an operation. Thus the alternative word 'useless' has some merit. We will, however, continue to use the word 'unfeasible' since it has become entrenched in the literature, and the false nature of the implication of a pathway is understood by those that use the groups in molecular spectroscopy. The MS group of a molecule is obtained by deleting all elements from the CNPI group of the molecule that are unfeasible; the elements in the MS group are said to be feasible. An unfeasible element is one that interconverts numbered equilibrium versions of the molecule when these versions are separated by an insuperable barrier in the potential energy surface; an insuperable barrier is one that does not allow observable tunneling to occur through it on the time scale of the experiment being performed.

Tunneling may not be observable if we use a low resolution experiment yet it may occur if we use a high resolution experiment. The MS group that we use to analyze the results will then be different for the two cases since the elements associated with the tunneling are feasible in the latter (high resolution) case but not in the former (low resolution) case. If splittings from inversion tunneling were observed in methyluoride (perhaps in highly excited vibrational states), and we wished to symmetry label the split levels, the MS group of the molecule would become equal to its CNPI group since all elements of it would be feasible.

In the same way that to set up the point group for a molecule we need to know the equilibrium geometry, so to set up the MS group we need to know the equilibrium geometry and the situation with regard to vibrational tunneling.

We will use the terms 'feasible' and 'unfeasible' in two different ways. The first way is as discussed above when determining which permutations in the CNPI group should be retained in the MS group. The second way is rather closely related and it concerns the labeling of tunneling motions through barriers in the potential function of a molecule. If such tunneling motions (degenerate rearrangements) produce observable splittings or shifts then such motions will be called feasible, and if they do not produce observable splittings or shifts then they will be called unfeasible. We call geometries that are accessible along feasible tunneling paths 'accessible geometries'; thus the planar geometry of ammonia is accessible. All these definitions, of course, depend on there solution of the experiment being used to study the molecule in question.

For the ethylene molecule the CNPI group has 96 elements and is the direct product $S^{(H)}_4 \otimes S^{©}_2 \otimes \varepsilon$. The CNPI group of one of the versions (a), (b), (c), or (d) shown in Fig. 3 can be used as the MS group of ethylene, and it consists of the eight elements given in Eq. (15). If we wish to use one of the numbered versions (e), (f), (g), or (h) in Fig. 3, in the vibrational problem, then the MS group we use is

$$\{E;\ (13)(24);\ (12)(34)(56);\ (14)(23)(56);$$

$$E^*;\ (13)(24)^*;\ (12)(34)(56)^*;\ (14)(23)(56)^*; \qquad (2)$$

whereas if we want to use one of the numbered versions (i), (j) (k) or (l) in Fig. 3 the appropriate MS group is

$$\begin{pmatrix} \{E;\ (14)(23);\ (12)(34)(56);\ (13)(24)(56); \\ E^*;\ (14)(23)^*;\ (12)(34)(56)^*;\ (13)(24)(56)^*; \end{pmatrix} \qquad (3)$$

To solve the vibrational problem we use one version of the ethylene molecule, and the MS group is the one appropriate for that version. Regardless of which version we chose we obtain an identical energy level pattern and symmetry labeling. Note that for a planar molecule the inversion operation E^* is always feasible, and the MS group is the direct product of the group of feasible permutations and the inversion group $\varepsilon = \{E;\ E^*\}$.

The MS group is thus denied for a particular electronic state (and numbered version) of a molecule with regard to the experimental observation or non observation of tunneling splittings. It therefore uses knowledge of the molecular structure and potential energy surface to obtain a symmetry group that is much smaller than the CNPI group (in most cases), but which provides a sufficient symmetry labeling of the observed levels. We return to the methane molecule with which we began this discussion. Omitting unfeasible elements from the CNPI group G_{48} gives the molecular symmetry group $T_d(M)$ having five irreducible representations. Using this group the rotation-vibration levels are labeled A_1, A_2, E, F_1 or F_2 and there is no redundant double labeling as obtained using G_{48}.

This labeling, in combination with the labeling obtained from the group K(spatial), is sufficient to understand the symmetry aspects of all intermolecular interactions and external perturbations with the exception of effects arising from inversion tunneling. If in some ultra-high resolution experiment inversion doublets were observed for methane then there would be ten rather than five distinguishable states (as far as nuclear permutation inversion symmetry were concerned) and to distinguish them one would need to use G_{48} since there would no longer be degeneracies between the states of opposite parity.

One systematic way to determine the MS group of a molecule is first to label the nuclei and then to write down the elements of the CNPI group of the molecule. The next stage is to draw all the distinct versions of the molecule at equilibrium (by permuting the labels on identical nuclei with and without inverting the molecule) and to form sets of these versions that one wishes to be connected by observable tunneling effects. To obtain the MS group it is then necessary to consider one particular set and to delete from the CNPI group of the molecule any elements that convert the versions in that set to the versions in other sets. Different sets may have different MS groups. In this approach one first writes down the versions and then determines the permutations and permutation inversions of one of the versions. By then determining the permutations that interconnect versions between which there is feasible tunneling one can rather quickly arrive at all the elements of the MS group. In the determination of larger MS groups it is useful to know that they can be written as the product of subgroups.

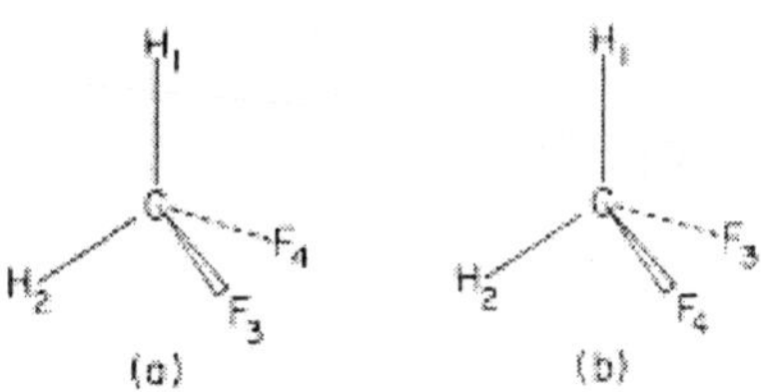

Fig. 5. The two versions of the equilibrium structure of a CH_2F_2 molecule.

Problem: Set up the MS groups of the following molecules in their ground electronic states: (i) CH_2F_2, (ii) HN_3, (iii) BF_3, (iv) NF_3, (v) CH_4, (vi) trans C(HF) CHF, and (vii) C_2H_2. Each of these molecules is such that either it does not possess more than one version of the equilibrium structure or, if it does, no observable tunneling between them occurs. Hence, they are all rigid molecules.

Answer. (i) H_2F_2. We choose to label the protons 1 and 2, and the uorine nuclei 3 and 4, so that the CNPI group is

$$\{E;\ (12);\ (34);\ (12)(34); E;\ (12);\ (34);\ (12)(34)g: \tag{4}$$

There are only two versions of the equilibrium structure as shown in Fig. 5, and no tunneling between these versions is observed. The version in Fig. (5a) has 2-3-4 clockwise when looking in the C $\rightarrow$ H_1 direction and the version in Fig. (5b) has them anticlockwise. The permutations (12) and (34) interconvert the versions ana are thus unfeasible, whereas (12)(34) is feasible since it does not interconvert the versions. The operation E^* interconverts the versions so that E^* is unfeasible. The operations (12)* or (34)* do not interconvert the versions and are both feasible. The operation (12) (34)* does interconvert the versions and is unfeasible. Thus the MS group of either version of CH_2F_2 is the group

$$\{E;\ (12)(34);\ (12)^*;\ (34)^*: \qquad (5)$$

(ii) HN_3. We choosé to label the nitrogen nuclei 1, 2, and 3 so that the elements of the CNPI group are

$$\left(\begin{array}{c}\{E;\ (12);\ (23);\ (13);\ (123);\ (132); E^*;\\ (12)^*;\ (23)^*;\ (13)^*;\ (123)^*;\ (132)^*:\end{array}\right) \qquad (6)$$

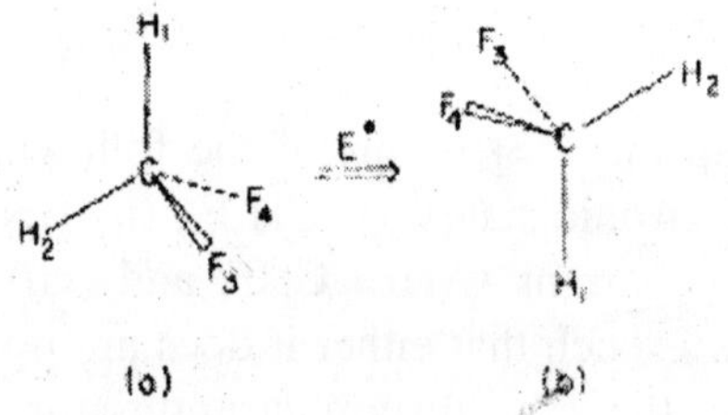

Fig. 6. The effect of the operation E on the version of CH_2F_2 given in Fig. 5a.*

The versions of the equilibrium structure are shown in Fig. 7. There is no observable tunneling between these versions. It is easy to see that all the permutations interconvert versions. For example, we determine that the version in Fig. (7a) ,is transformed as follows:

$$\left(\begin{array}{c}(12)(a) = (c);\ (23)(a) = (b);\ (13)(a) = (f);\\ (123)(a) = (d);\ \text{and}\ (132)(a) = (e):\end{array}\right) \qquad (7)$$

The inversion E^* is feasible since it does not interconvert versions but all permutations accompanied by the inversion are unfeasible. Thus the MS group of HN_3 is

$$\{E;\ E^*\} \qquad (8)$$

(iii) BF_3. We label the three uorine nuclei 1, 2, and 3 and hence the CNPI group is the same as that of HN_3 given in Eq. (6). However, in its equilibrium conguration BF3 is planar with three equal BF bonds, and as a result there are no different versions of the equilibrium structure; all operations of the CNPI group are feasible. Thus the MS group of BF_3 is the same as its CNPI group and there is no structural degeneracy. The H_2O molecule is another example of a simple molecule having no structural degeneracy for which the MS group is equal to the CNPI group.

(iv) NF_3. Labeling the three uorine nuclei 1, 2, and 3 we obtain the CNPI group of Eq. (6) as for HN_3 and BF_3. For this pyramidal molecule there are two versions of the equilibrium structure as shown in Fig. 8. To interconvert these versions we must invert the molecule, and inversion splittings are not observed. Deleting elements from the CNPI group that interconvert these versions we obtain the MS group of NF_3 as

$$\{E;\ (123);\ (132);\ (12)^*;\ (23)^*;\ (13)^*: \qquad (9)$$

(v) CH_4. Labeling the four protons 1, 2, 3, and 4 we can write down the CNPI group of the molecule as the direct product of the CNP group $S^{(H)}_4$ containing 4! = 24 elements and the inversion group ε; this has 48 elements. Any cyclic permutation of three protons is clearly feasible, and the inversion E^* is equally clearly unfeasible. From these three results we can rather quickly deduce the MS group of CH_4 by making use of the following rules:

(a) Any operation that is the product of a feasible operation and an unfeasible operation is itself unfeasible, and

(b) Any operation that is the product of two feasible operations is itself feasible.

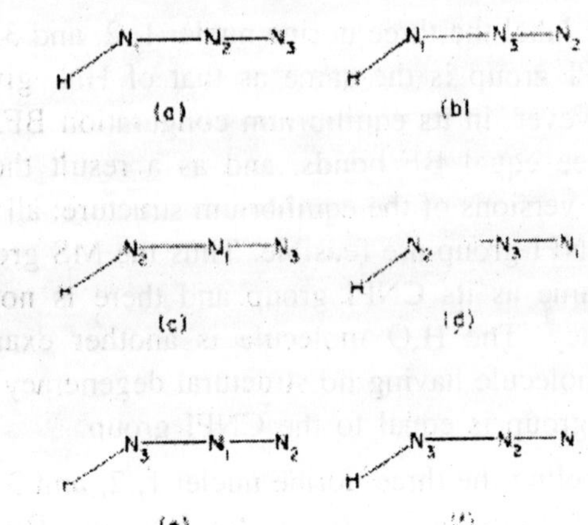

Fig. 7. The six versions of the equivalent equilibrium structure of an HN_3 molecule.

For an example of the use of (a) we can look at the product

$$(123)(34) = (1234) \tag{10}$$

The operation (34) converts the version *A* to version *C* (since it is an unfeasible operation) and (123) then sends *C* to *C* (since it is feasible), so the product of these two operations sends *A* to *C* and is unfeasible. For methane any cyclic permutation of all four protons is unfeasible. An example of rule (b) is provided by the product

$$(123)(234) = (12)(34) \tag{11}$$

The operation (234) converts *A* to *A* and then (123) converts *A* to *A*, so that the product converts *A* to A and is feasible. We can see this latter result another way by looking at this operation as the product (12)(34), and this is the successive application of two unfeasible operations: (34)*A* = *C*, and (12) *C* = *A*, so that the product sends *A* to *A* and is feasible. We can thus appreciate a third rule:

(c) For molecules with only two distinct numbered versions the product of two unfeasible operations is itself a feasible operation. For molecules having more than two distinct numbered versions the product of two unfeasible operations may, or may not, be feasible.

Fig. 8. The two versions of the equilibrium structure of an NF3 molecule.

For methane we deduce that all cyclic permutations of three protons and all permutations consisting of two successive pair transpositions are feasible whereas all pair transpositions, and cyclic permutations of all four protons are unfeasible.

Since E^* is unfeasible we can use rules (a) and (c) to deduce that the product of a pair transposition and E^* or the product of a cyclic permutation of all four protons and E^* is feasible. The MS group of methane thus consists of the following 24 elements:

E	(123)	(12)(34)	(1234)	(12)	
	(132)	(13)(24)	(1243)	(13)	
	(124)	(14)(23)	(1324)	(14)	
	(142)		(1342)	(23)	
	(134)		(1423)	(24)	(12)
	(143)		(1432)	(34)	
	(234)				
	(243)				

(vi) trans C(HF)CHF. Numbering the protons 1 and 2, the carbon nuclei 3 and 4, and the uorine nuclei 5 and 6, we obtain the CNPI group as the direct product

$$\left(\begin{array}{c} \{E;\ (12);\ (34);\ (56);\ (12)(34); \\ (12)(56);\ (34)(56);\ (12)(34)(56)\} \otimes \varepsilon \end{array} \right) \quad (13)$$

All four versions of the equilibrium structure are given in Fig. 9. It is important to realize that a structure such as given in Fig. 10 is not symmetrically equivalent to those in Fig. 9 since it is not obtained from any of them by relabeling identical nuclei with or without inverting the molecule.

The structure in Fig. 10 is the cis structure of the molecule and is related to the trans structure by twisting the molecule over a potential barrier. The distinction between structure and version is discussed by Bone, Rowlands, Handy and Stone. In this molecule the ground electronic state potential energy surface has four versions for the trans structure and four versions for the cis structure at a different energy from that of the trans structure.

The shape of the potential energy surface around the cis and trans minima will be different and the vibrational energy level patterns of the cis and trans structures will be different. For trans $C_2H_2F_2$ the feasible elements are easy to determine and the MS group consists of

$$\{E;\ (12)(34)(56);\ E^*;\ (12)(34)(56)^*\} \quad (14)$$

The MS group of cis $C_2H_2F_2$ is the same as that of trans $C_2H_2F_2$ and if cis-trans tunneling occurs the MS group is still the same. The tunneling is not between versions of the equilibrium structure, so although it makes the molecule non rigid it does not cause a splitting of any structural degeneracy and it does not enlarge the MS group.

(vii) C_2H_2. In the acetylene molecule HCCH we number the protons 1 and 2 and the carbon nuclei 3 and 4; the CNPI group is the group of eight elements given in Eq. (4). The operations (34); (12); (34), and (12) are unfeasible and the MS group is

(15)

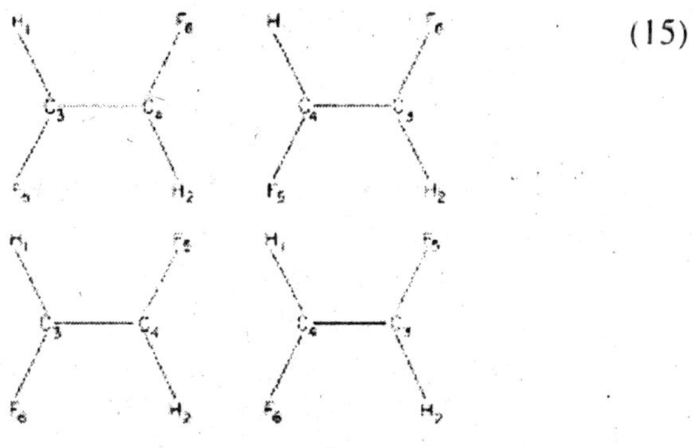

Fig. 9. The four versions of the equilibrium (trans) structure of the C(HF)CHF molecule.

For any symmetrical linear molecule the MS group consists of the four elements E, (p), E, and (p), where (p) is the simultaneous transposition of all pairs of identical nuclei symmetrically located about the molecular mid-point. For any unsymmetrical linear molecule (such as HCN) the MS group is $\{E, E^*)$.

Problem 2. Set up the MS groups of the following non rigid molecules in their ground electronic states: (i) NH_3 allowing for inversion tunneling; (ii) C_2H_4 and (iii) H_2O_2 allowing for torsional tunneling in each case. Answer. (i) NH_3 with inversion tunneling. Numbering the protons 1, 2, and 3 the CNPI group is as for HN_3 given in Eq. (6), and there are two versions of the equilibrium structure as for NF_3 (Fig. 8).

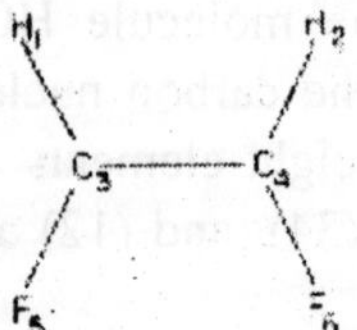

Fig. 10. A numbered version of the cis structure of the C(HF)CHF molecule.

Allowing for the effects of inversion tunneling in setting up the MS group we must include as feasible elements permutations and permutation inversions that interconvert versions across the inversion barrier. Thus the MS group of inverting NH_3 is the same as its CNPI group and this is the same as the MS group of planar BF_3. (ii) C_2H_4 with torsional tunneling. The CNPI group of ethylene is the direct product $S^{(H)}_4 \otimes S_2^{(C)} \otimes \varepsilon$ with 96 elements as discussed before. Grouping these into sets within which the versions are connected by torsional tunneling we obtain the six sets [(a),(b)], [(c),(d)], [(e),(f)], [(g),(h)], [(i),(j)], and [(k),(l)] from Fig. 3. Let us consider the set [(a),(b)].

Examples of elements that convert the versions in this set into versions in other sets are (13), (56), (24), and (234), and these are hence unfeasible elements of the CNPI group for the set [(a),(b)]. Elements such as (12) and (34) are feasible, and combining these with the elements of the MS group of non torsionally tunneling ethylene in the versions (a) and (b) we obtain the MS group of the torsionally tunneling versions in set [(a), (b)] of Fig. 3 as

$$E \begin{pmatrix} (12) & (12)(34) & (13)(24)(56) & (1324)(56) \\ (34) & & (14)(23)(56) & (1423)(56) \end{pmatrix} \otimes \varepsilon. \quad (16)$$

The MS group for the set [(c), (d)] is the same as this. The MS group for the set [(e), (f)] or the set [(g), (h)] is the group

$$E \quad \begin{pmatrix} (13) & (13)(24) & (12)(34)(56) & (1234)(56) \\ (24) & & (14)(23)(56) & (1432)(56) \end{pmatrix} \otimes \varepsilon. \quad (17)$$

and the MS group for the set [(i), (j)] or the set [(k), (l)] is the group

$$E \quad \begin{pmatrix} (14) & (14)(23) & (12)(34)(56) & (1243)(56) \\ (23) & & (13)(24)(56) & (1342)(56) \end{pmatrix} \quad (18)$$

Since we only consider a single set of versions of the equilibrium structure connected by allowed tunneling, and neglect the others, in a vibrational analysis, the fact that there are different MS groups poses no problems as long as we are careful to use the one appropriate to the particular set considered. These groups lead to identical energy level labelings of the appropriate sets.

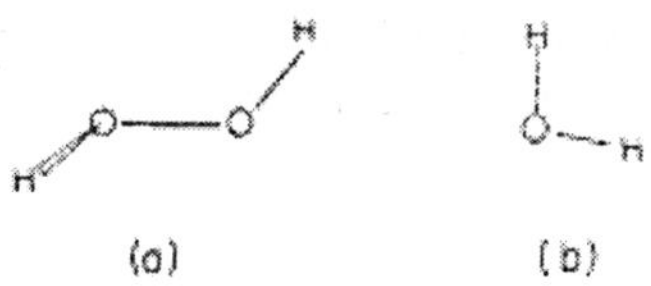

Fig. 11. The H_2O_2 molecule in its equilibrium configuration; (a) is a view from the side and (b) is an end on view.

(iii) H_2O_2 with torsional tunneling. The equilibrium structure of H_2O_2 in its ground electronic state is nonplanar with a dihedral angle of approximately 120, as shown in Fig. 11. Labeling the protons 1 and 2, and the oxygen nuclei 3 and 4, we can label the nuclei of the conguration of Fig. 11 in the two ways shown in (a) and (b) of Fig. 12. A nonsuperposable version of the equilibrium conguration is obtained from that of Fig. 11 by inverting it, and the two ways of labeling this conguration are shown in (c) and (d) of Fig. 12.

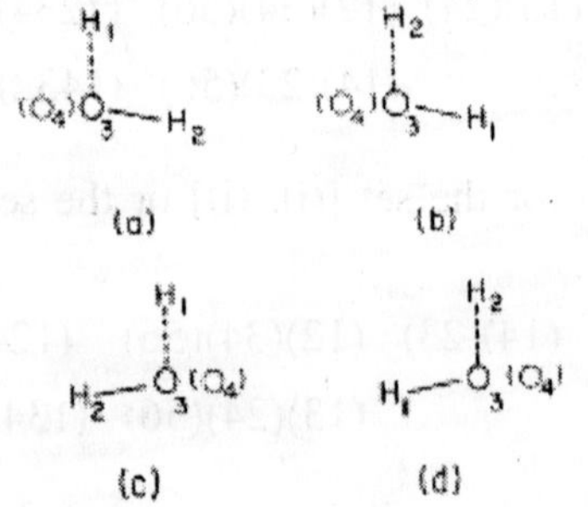

Fig. 12. The four versions of the equilibrium structure of an H_2O_2 molecule shown viewed end on. The oxygen atom O_4 is behind the oxygen atom O_3 in the four figures.

There are thus four versions of the equilibrium structure of H_2O_2 in its ground electronic state. If there were no observable tunneling between these structures each vibrational level would have a fourfold structural degeneracy and the MS group would be

$$\{E;\ (12)(34)\} \tag{19}$$

Allowing for torsional tunneling the interconversions (a) $\leftrightarrow$ (c) and (b) $\leftrightarrow$ (d) of Fig. 12 are feasible, and we have two sets of structures, [(a),(c)] and [(b), (d)]. The MS group for either of these sets is

$$\{E;\ (12)(34); E^*;\ (12)(34)^*\} \tag{20}$$

In H_2O_2 torsional tunneling is observed and the MS group is as given in Eq. (20). The H_2S_2 molecule has a similar equilibrium structure to that of H_2O_2, with a dihedral angle of about 90. Except in ultrahigh resolution Lamb-dip experiments H_2S_2 has no observable torsional tunneling in its ground state, and thus for normal spectroscopic work on this state its MS group is as given in Eq. (19); no permutation inversion elements are present.

A molecule having no permutation inversion elements in its MS group has optically active versions and H_2S_2 is a simple example of this.

Weakly Bound Cluster Molcecules

Beginning with the pioneering work of Harry Welsh and his colleagues at the University of Toronto in the 60's and early 70's the high resolution spectroscopic study of weakly bound molecular clusters, often called Vander Waals molecules, has grown into a topic of widespread interest and importance. We have a weakly bound molecular cluster as a complex formed in the gas phase (or in a molecular beam) between two or more stable neutral monomers; the monomers can be atoms or molecules. We will call the monomer units within the cluster moieties so that we are free to use the word molecule for the cluster itself without ambiguity. By far the most widely studied molecules of this type are dimers: there are homogeneous dimers (having identical moieties) such as $(H_2)_2$, $(HF)_2$, $(H_2O)_2$, and $(NH_3)_2$, and heterogeneous dimers (having dissimilar moieties) such as Ar the relevant tunneling splittings.

Minima other than those at the global minimum of the potential function (i.e., other than at versions of the equilibrium structure) may also be important for the analysis of the spectrum. For a weakly bound cluster molecule it is frequently necessary to consider the subgroup of the CNPI group obtained by deleting only elements that involve unfeasible contortions of each of the moieties; we call this group the full cluster tunneling (FCT) group. This group is obtained from the CNPI group of the molecule by deleting all elements that correspond to degenerate rearrangements involving a path identical to one that would be unfeasible in an isolated moiety. Thus in the methane-argon dimer the FCT group would be the same as the MS group of the methane moiety and E would be unfeasible; this group happens to be the MS group of the dimer. However, sometimes the MS group of the cluster molecule is a subgroup of the FCT group.

The first example of a weakly bound cluster molecule that we consider is the hydrogen dimer $(H_2)_2$. We label the protons H_1H_2–H_3H_4. This is an example of a dimer in which the vander Waals bond (which we can draw for this molecule as connecting the mid-points of the moieties) is only weakly an isotropic so that all angular orientations of each moiety with respect to the bond are feasible. Thus all elements of the CNPI group are feasible except those such as (13) that involve the breaking and re-forming of the covalent HH bonds of the moieties, and the MS group is the FCT group; this is given by

$$E \quad \begin{pmatrix} (12) & (12)(34) & (13)(24) & (1324) \\ (34) & & (14)(23) & (1423) \end{pmatrix} \otimes \varepsilon. \quad (21)$$

For the isotopomer $(HD)_2$ we label the nuclei H_1D_2 – H_3D_4. Since H and D nuclei are not identical the permutations (12) and (34) cannot be present in the symmetry group and so the MS group is

$$\{E;\ (13)(24); E^*;\ (13)(24^*)\} \qquad (22)$$

Such a weakly anisotropic intermoiety bond as is present in the hydrogen dimer usually only occurs for bonds to rare gas or H_2 moieties. A more typical example is the HF dimer already mentioned above. For the HF dimer the two versions of the equilibrium structure are approximately as shown in Figs. 13a and 13c, and the structure at the saddle point transition state between them is as shown in Fig. 13b. The tunneling motion between these versions involves the breaking and forming of a van der Waals bond at the transition state. Dyke, Howard and Klemperer (1972) discovered that this tunneling is feasible (producing a splitting of 0.65 cm^{-1} in a very important microwave spectrocopy study, and this means that the MS group is the FCT group given in Eq. (22) above.

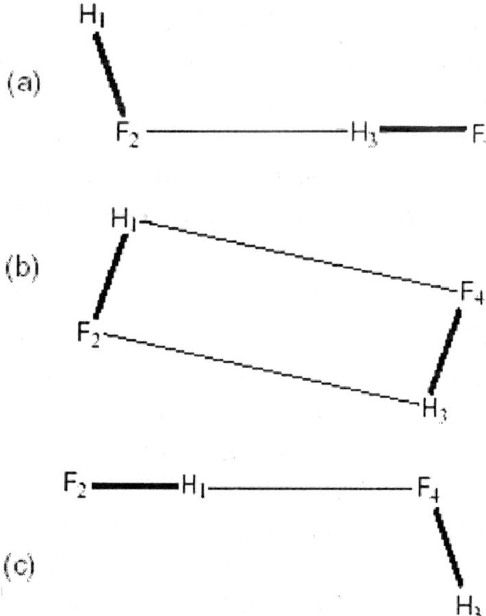

Fig. 13. An HF dimer molecule in its two different versions of the equilibrium structure, (a) and (c), which can be interconverted by a tunneling through the symmetrical transition state shown in (b).

The water dimer is an example with more than one feasible degenerate rearарangement path. Numbering the protons 1 and 2, and the oxygen nucleus 5, on one water moiety, and the protons 3 and 4, and the oxygen nucleus 6, on the other, then the FCT group is essentially the same as that for the hydrogen dimer:

$$E \begin{pmatrix} (12) & (12)(34) & (13)(24)(56) & (1324)(56) \\ (34) & & (14)(23)(56) & (1423)(56) \end{pmatrix} \otimes \varepsilon. \quad (23)$$

This group is the same as the MS group G_{16} of ethylene. If tunnelings between all versions of the equilibrium structure (except those involving the breaking of the covalent OH bonds in the moieties) are feasible then this is the MS group. However, to see if this is the case, and to use the group, we need to study the situation further with the help of a theoretical study of the effects of the various tunneling processes and some ab initio calculations. The shape of the equilibrium structure is determined to be as shown approximately in Fig. 14; this is very much like the HF dimer in that there is a hydrogen bond between the moieties with one moiety being the hydrogen bond donor (one of its protons forms the hydrogen bond) and the other moiety being the hydrogen bond acceptor.

By studying Fig. 14 we determine that there are eight versions of the equilibrium structure (ignoring the possibility of breaking the covalent OH bonds in the moieties): each of the four protons in turn can be the hydrogen bonding proton, and for each of these four choices there are two choices for the labeling of the protons in the acceptor moiety. The eight versions of the equilibrium structure are shown in Fig. 15, and we label these versions (1) to (8) as in Fig. 4 of Coudert, Lovas, Suenram and Hougen. We have not drawn all these versions the same way around in order that the degenerate rearrangements that closely parallel the one occurring in the HF dimer should be apparent.

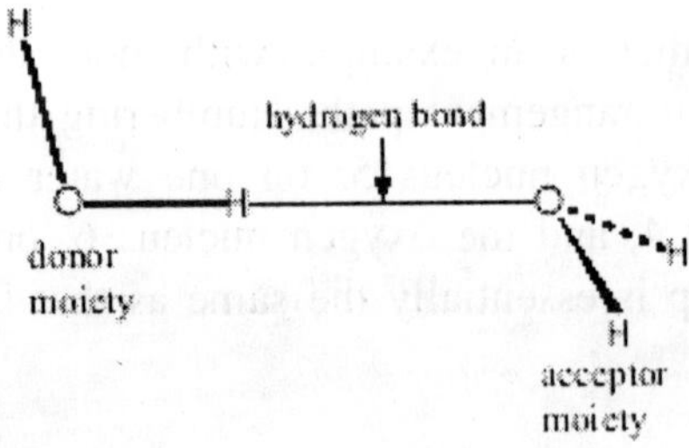

Fig. 14. The equilibrium structure of the water dimer

The four solid double headed arrows [such as that which connects (1) and (4)] show the degenerate rearrangement with the lowest hindering barrier, and this is essentially a rotation of the acceptor moiety about its two-fold axis. This has a barrier of about 200 cm^{-1} and is called the acceptor-tunneling path. The eight dashed double headed arrows [such as that which connects (1) and (5)] show the degenerate rearrangement that parallels that in the HF dimer. In this rearrangement the acceptor moiety rotates into the hydrogen bond while, in a geared fashion, the donor moiety rotates out. This has a barrier of about 300 cm^{-1} and is called the donor-acceptor interchange path.

The four broad double headed arrows [such as that which connects (1) with (2)] shows the degenerate rearrangement that involves the concerted rotation of the donor moiety (to exchange which of its protons participates in the hydrogen bond) and inversion at the O atom of the acceptor moiety. This has a barrier of about 650 cm^{-1} and is called the donor-tunneling or bifurcation path. We could imagine different situations depending on the feasibility or not of the various degenerate rearrangements, and depending on the experimental resolution with which the rotation-vibration energy levels were measured. If all degenerate rearrangements were unfeasible the water dimer would be a rigid molecule and the MS group of versions (1) to (4) would be $\{E,(34^*)$, whereas that of versions (5) to (8) would be $\{E,(12^*)\}$.

Each of the eight versions would have the same rotation-vibration energy levels. Ignoring the structural degeneracy resulting from bond breaking in the moieties each level would have an eight-fold structural degeneracy. If the acceptor-tunneling path (solid arrows) were considered feasible but the other degenerate rearrangements still unfeasible then (34) would become feasible for versions (1) through (4), and (12) would become feasible for versions (5) through (8). The MS group of versions [(1),(4)] and [(2),(3)] would become $\{E;\ (34),\ E^*,\ (34)^*\}$ and the MS {Eg group of versions [(5),(8)] and [(6),(7)] would become $\{E;\ (12)\ \{E^*\ (12)^*\}$. The molecular energy levels would now have a resolved splitting into two levels each of which would have a four-fold structural degeneracy remaining because of the unfeasibility of the other degenerate rearrangements. If the donor-acceptor interchange path were also considered to be feasible then all eight minima would be accessible to each other and in this circumstance the MS group would be the FCT group given in Eq. (23).

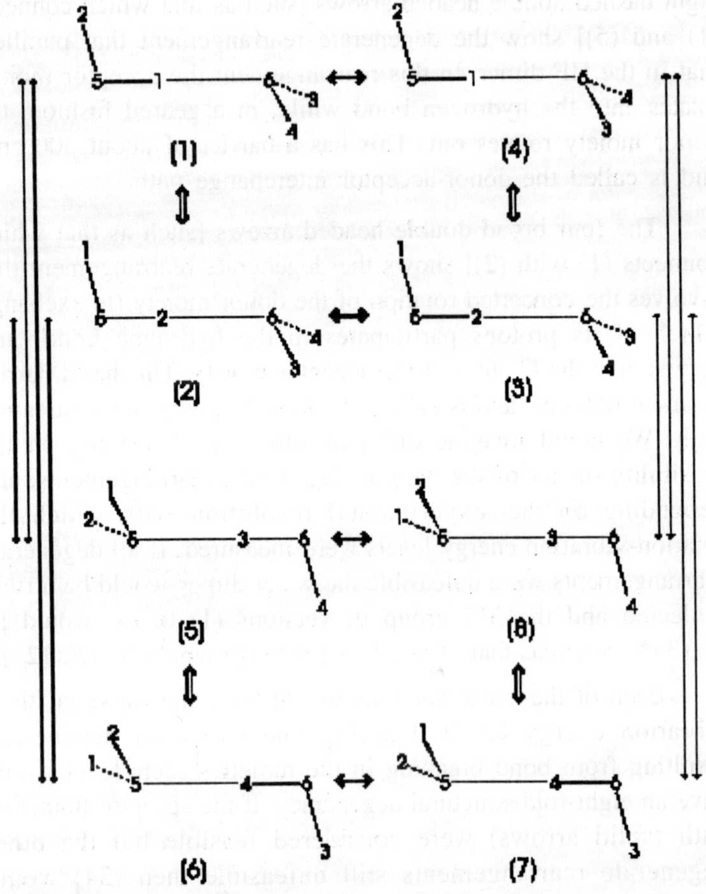

Fig. 15. The eight different versions of the equilibrium structure of a water dimer molecule, (1) through (8), which can be interconverted by the various tunneling rearrangements indicated by the double headed arrows.

No further splittings would be possible although the tunnelings caused by other paths [such as the donor-tunneling path] would modify the splittings caused by the acceptor-tunneling and donor-acceptor interchange paths just considered. The experimental situation in this molecule is that with the normal resolution of microwave spectroscopy all these possible splittings are resolved and the MS group that one needs to label the distinguishable energy levels is the FCT group given in Eq. (23). Although in the light of the experimental observations and the ab initio calculations it is not the case, it would be instructive for the reader to determine the MS group in the event that the donor-tunneling and acceptor-tunneling paths were feasible but the donor-acceptor interchange were not. Permutations that interchange the moieties would not be feasible in this case.

As an example of a dimer having a rather large number of versions of its equilibrium structure, and having also more than one degenerate rearrangement path, we consider the benzene-water dimer. We base this discussion on the theoretical calculations of Gregory and Clary. We label the pairs of bonded CH nuclei of the benzene moiety 1, 2, ..., 6 cyclically, and the protons on the water moiety 7 and 8. The equilibrium structure of the dimer can be approximately described as having one OH bond of the water moiety perpendicular to the plane of the benzene moiety with the H atom of that bond pointing at the center of the ring; the benzene ring acting as a hydrogen bond acceptor. There are 24 versions of the equilibrium structure of this molecule.

For the benzene-water dimer the internal rotation about the hydrogen bond has a very low six-fold barrier, and thus this internal rotation tunneling will produce a large splitting; this means that permutations such as (123456), (135)(246) and (14)(25)(36) are feasible. This rearrangement connects six versions of the equilibrium structure, and there are four distinct sets of six such versions depending on which H forms the hydrogen bond, and on which face of the benzene molecule accepts this hydrogen bond. There is an intermediate barrier to the tunneling motion that involves the exchange of the water moiety hydrogen bonds; in this motion there is a rotation of the H_2O so that the hydrogen atoms involved in the hydrogen bond are exchanged. Considering this motion to be feasible means that the permutation (78) is feasible.

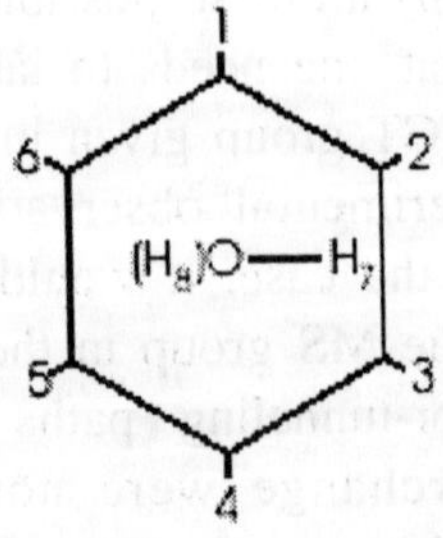

Fig. 16. A version of the equilibrium structure of the benzene-water dimer.

A third degenerate rearrangement that is possible on this surface is the reverse ring tunneling that involves a concerted rotation of both moieties so that the other face of the benzene ring is involved in the hydrogen bond. This motion is hindered by a high barrier and can be considered unfeasible. Thus permutations such as (12)(36)(45) and (15)(24), and the inversion E^*, are unfeasible. In this circumstance the 24 versions of the equilibrium conguration divide into two distinct sets of 12; within each set of 12 tunneling is feasible, but tunneling is unfeasible between the minima of the two sets. With these considerations the MS group of the benzene-water dimer is of order 24. This group, G_{24}, is the direct product of the group $C_{6v}(M)$ with elements

$$E\begin{pmatrix} (123456) & (135)(246) & (14)(25)(36) & (26)(35) & (14)(23)(56)^* \\ (165432) & (153)(264) & & (31)(46) & (25)(34)(61)^* \\ & & & (42)(51) & (36)(45)(12)^* \end{pmatrix} \quad (24)$$

and the group

$$\{E^*\ (78)\} \quad (25)$$

G_{24} is half the size of the FCT group which would be given by the direct product $G_{24} \otimes \varepsilon$. The FCT group would be the MS group if the reverse ring tunneling were also feasible. If we chose to consider the hydrogen exchange tunneling and the reverse ring tunneling as both being unfeasible then the MS group would simply be the group $C_{6v}(M)$ given in Eq. (24). This is the MS group of the benzene-argon dimer. If there were no feasible degenerate rearrangements, i.e., if the molecule were rigid, then the MS group of the version of the benzene-water cluster molecule given in Fig. 16 would be the $C_s(M)$ group

$$\{E^* \ (14)(23)(56)^*\} \tag{26}$$

The MS Group for Levels of More than one Electronic State

So far in this chapter we have set up the MS group in order to classify the rotation-vibration levels of a single electronic state of a molecule. The electronic wavefunctions of a molecule depend on the nuclear coordinates, and for a particular electronic state, we can classify the electronic wavefunction, and hence rovibronic wavefunctions, in the MS group of that electronic state. It is sometimes necessary to consider rovibronic levels of more than one electronic state in a given problem; for example, we might want to consider interactions between rovibronic levels belonging to different electronic states or to consider electric dipole transitions between electronic states. In these circumstances we must generalize our definition of the MS group so that it will be appropriate for classifying simultaneously the rovibronic levels of more than one electronic state.

Clearly, just as for the purposes of classifying the levels of a single electronic state, we could use the CNPI group for classifying the levels of all electronic states of a molecule. However, we wish to use the smallest group necessary for a sufficient symmetry labeling of the levels, and this group is the MS group. The definition of the MS group given earlier for one electronic state can be applied when we wish to study simultaneously more than one electronic state by generalizing the concept of feasibility. The generalization is to say that an unfeasible element of the CNPI group interconverts versions of the equilibrium structure of one of the electronic states being considered when these structures are separated by an insuperable potential energy barrier and when they cannot be interconverted by transitions to and from any of the other electronic states being considered.

An example of the application of this extended definition is to the determination of the MS group appropriate in the study of the V ←N electronic transition in ethylene. The ethylene molecule is planar in the equilibrium structure of the N ground electronic state, as we have discussed above, but in the V excited electronic state it is nonplanar at equilibrium with one CH_2 twisted at 90 relative to the other about the C-C bond. The MS group of the numbered ground state version in Fig. 3a is given in Eq. (15). Within the ground state this numbered version has negligible tunneling into the other numbered versions. However, by transitions to and from the V state the version in Fig. 3a can be converted into the version in Fig. 3b. Thus the elements (12) and (34) are feasible here and are in the MS group for use in simultaneously classifying the levels of the N and V electronic states of ethylene: these elements are unfeasible for either of the electronic states taken by themselves.

Hence the MS group for use in simultaneously classifying the levels of the N and V states is the group given in Eq. (16). The excited R electronic state of ethylene is nonplanar and in a study of the R ← X band system the same MS group is required. The symmetry of ethylene, and the use of that symmetry in understanding the *V* -N resonant Raman spectrum.

The CNPI group of a molecule can be used to label its energy states with irreducible representation labels. However, for a molecule in an electronic state in which there are two or more versions of the equilibrium structure, between which tunneling produces no observable splittings, this classification scheme is more detailed than necessary. To symmetry label the energy levels as much as is necessary, that is in order to distinguish between energy levels that are distinguishable experimentally, it is sufficient to use a particular subgroup of the CNPI group. This subgroup, called the molecular symmetry (MS) group, is obtained from the CNPI group by deleting all unfeasible elements. If we are only interested in considering a single electronic state of a molecule then an unfeasible element is one that interconverts versions of the equilibrium structure that are separated by an insuperable barrier in the potential energy function (on the time scale of the experiment); feasible elements maintain an equilibrium version as itself or interconvert versions between which tunneling (such as by torsion or inversion) produces observable splittings.

Rigid molecules are dened to be molecules for which all such tunneling rearrangements are unfeasible; nonrigid molecules have at least one feasible tunneling rearrangement. Weakly bound cluster molecules (often called van der Waals molecules) form a special class, and such molecules are treated separately. If we are interested in considering two or more electronic states of a molecule at once, because we wish to study interactions or transitions between them, then the definition of a feasible element must be extended to include elements that connect versions that can be interconverted by transitions to and from any of the other electronic states being considered.

6

Vibrational Spectroscopy

Symmetry operations are acts of rotating and reflecting a molecule or figure. Symmetry elements are the symmetry operations that can be done on a molecule to leave it indistinguishable from the way it was started. If we say that a molecule has symmetry C_4 about a given axis, we mean that 4 equal rotations (of 90 degrees each) about that axis will get us back to where we started, and that each of the four rotations will leave the molecule in a position identical to its starting position.

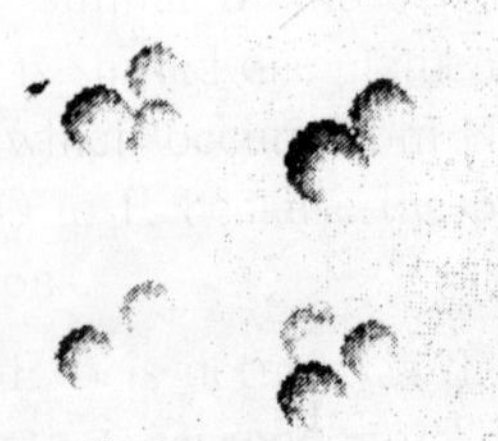

Notice also the reflection plane (reflection planes are denoted with the Greek letter σ, or sigma) that starts out in the plane of the molecule. Note that in these interactive figures, the bonds are not shown. Only the atoms are there. You should learn to recognize the variuos ways that chemists represent three-dimensional structures on two-dimensional paper. The figure shows that the four carbon atoms approximately define a planc; the hydrogens are on either side of the plane. Even though carbons are singly-bonded to each other, there is not free rotation this carbon-carbon bond because there is not room for the hydrogens to squeeze past each other though the ring.

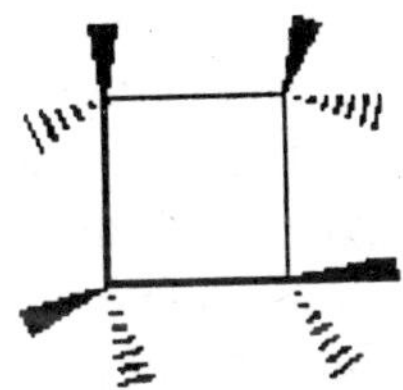

If a molecule has symmetry C_3 about some axis, we mean that three equal rotations (of 120 degrees each) about that axis will do the same thing. Methane, a tetrahedral molecule, is shown below. Methane also has 6 reflection planes. Each reflection plane is defined by the central carbon and two of the hydrogens.

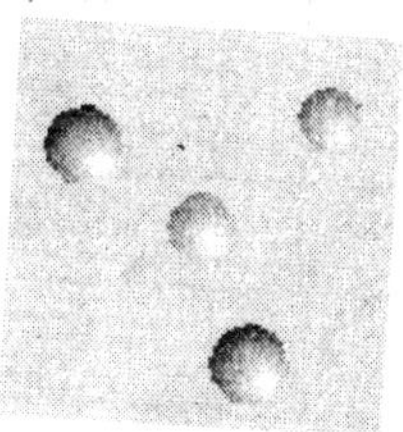

Here are some alternate representations of methane. The first shows explicitly that the four lobes of the tetrahedron point toward the corners of a cube.

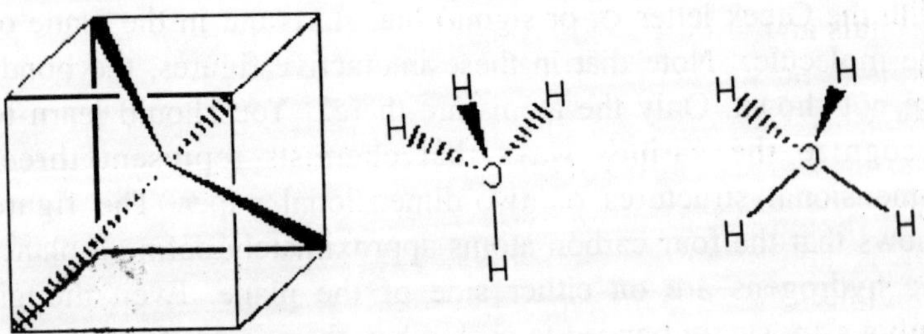

All symmetry operations are some combination of rotations and reflections. However, sometimes we do what is called an "improper rotation" in which we rotate the molecule and then reflect it through a plane perpendicular to the axis of the rotation. So, for example, an S_4 is just like a C_4 except that after rotating the molecule 90 degrees about some axis, we reflect it through a plane perpendicular to that axis. The methane molecule above has an S_4. The Official Chem 32 Theme Molecule, allene, also has an S_4. Allene is pictured below. Notice that if you were to replace just one of the hydrogens with a heteroatom such as chlorine, the S_4 symmetry is destroyed.

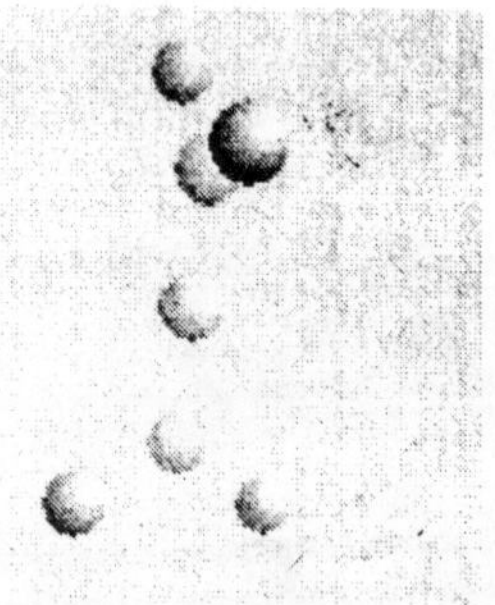

One of the reasons allene is so popular in this class is that it is not only interesting from a group theoretical perspective, but also from a bonding perspective. The central carbon is sp hybridized. Its two unhybridized *p* orbitals are required by quantum mechanical laws to be perpendicular to the *sp* hybrid orbitals and to each other. The consequence is that the *sp* hybrid orbitals on the central carbon form sigma bonds with the two outer carbons, one *pi* bond is formed in the plane of the page, and the other pi bond is perpendicular to the plane of the page. Below is a projection down the carbon-carbon-carbon bonds of allene and a picture of the p orbitals that are involved in its bonding scheme. This type of system, in which two adjacent pi bonds are formed from perpendicular sets of orbitals, is called a cumulated system and is somewhat unstable (although allenes are found in nature). Later we will contrast such systems with their more stable, conjugated counterparts.

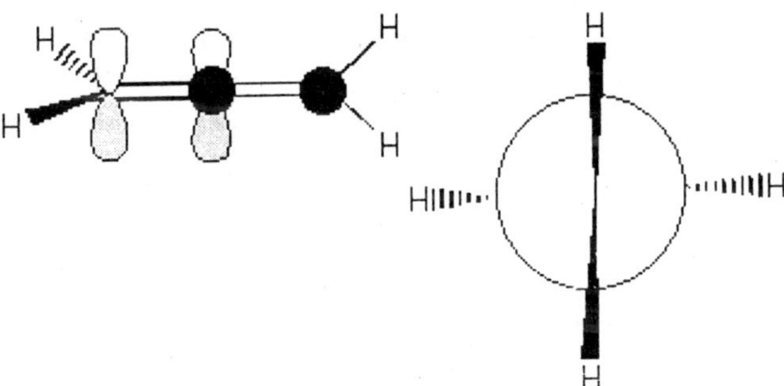

Chemists often talk about "inverting" a molecule (or orbital) or of its having even or odd "inversion symmetry." A molecule that is even to inversion has the property that if you take every point in the molecule, pass it through the molecule's center, and bring it out the same distance on the other side, the molecule will be unchanged. An orbital that is odd to inversion has the property that if you take every point of that orbital, pass it through the orbital's center, and bring it out the same distance on the other side, the sign of the orbital's wavefunction will be inverted (i.e. its shading will be flipped). "Inverting" a molecule is the same as doing an S_2 improper rotation. Notice that a simple reflection is also the same as doing an S_1 improper rotation.

These two facts become especially useful when we talk about chirality. Chiral molecules are defined as molecules that have no improper rotation axes. So if a molecule has reflection or inversion symmetry in any of its conformations, then you immediately can tell that the molecule is not chiral. All of the moveable molecules featured above do have reflection or inversion symmetry, so none of them are chiral. You may have heard of chiral molecules before and been told that a chiral molecule is a molecule having a carbon with four different substituents (such a carbon is often called a "chiral center").

However, since the carbon-carbon bond is a single bond, there is free rotation about that bond. If a molecule has reflection or inversion symmetry in any of its conformations, it is achiral. When we look down the carbon-carbon bond as in the bottom figures (these are called "Newman projections") the inversion symmetry in the "staggered conformer" and the mirror reflection plane in the "eclipsed conformer" are especially evident.

This is a molecule that most certainly does not possess a carbon with four different substituents—each carbon has a double bond!! However, it does not possess reflection symmetry, inversion symmetry, or any other higher-order improper rotation axes either. This is a chiral molecule without a chiral center.

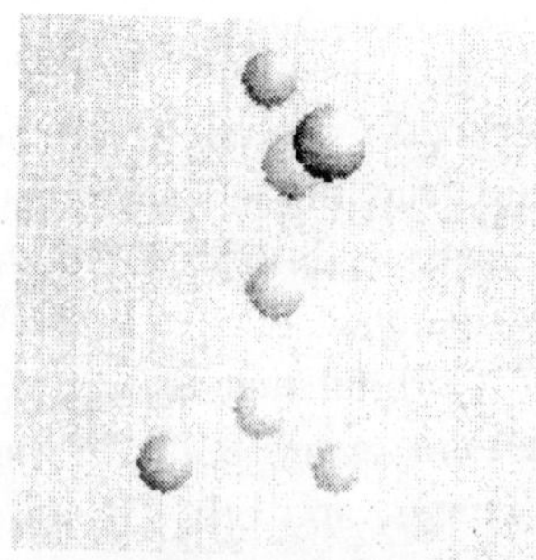

(It is worth noting on the side that just because you do not see any reflection planes or inversion centers, cannot immediately say a molecule is chiral. There might be a higher-order improper rotation axis that renders the molecule achiral. This is extremely rare) Why all this fuss about whether a molecule is "chiral" or not? Chiral molecules have the property that they are "handed." Just as you only shake hands right-hand to left-hand, chiral molecules in your body can only "shake hands" (i.e. interact) with certain other chiral molecules. This has far-reaching biological implications that we will explore. For example, are chiral proteins that are evolutionarily primed to recognize very specific molecules in very specific chiral configurations. We will talk about this extensively later; the point of this section is simply to make sure that you will be able to recognize chiral molecules when they come up. Also, chiral molecules have the physically interesting property that they rotate plane polarized light. In fact, this is how they were first identified.

Louis Pasteur, a French chemist, was puzzled by the fact that crystals that formed on some wine bottles rotated plane polarized light in opposite directions. By coincidence, the crystals themselves were handed too, and Pasteur was able physically to separate the ones that rotated light clockwise from the ones that rotated it counterclockwise. Jean Baptiste Biot, the deal of French optical rotation studies, was skeptical and made Pasteur do it in front of him before he believed the result. It was a great moment in the history of chemistry.

We can symmetrically classify a molecule or orbital in terms of its point group. Molecules in the same point group have the same symmetry elements. Professor Zare has provided a decision tree that you can use to decide what point group to put a molecule into. Another option is to internalize the following classification scheme. There are four sets of point groups. The first is for molecules with low symmetry: C_1 (only E), C_s (only Σ and sigma), and C_i (only E and I). The second is for molecules with high symmetry: T_d, (tetrahedral), O_h (octahedral), and I_h (icosohedral). If a molecule does not fall into the low symmetry groups or the high symmetry groups, then it is in either one of the C_n groups or the D_n groups.

The difference is that a molecule in a D_n group has n C_2 axes perpendicular to the principle axis, where n is the order of the principle axis. The subscripts are as follows: if there is a sigma *h* (perpendicular to the principle rotation axis), add an *h*; if there is no sigma *h* but a sigma *v* (containing the principle rotation axis), add a *v*; if there are no reflection planes, do not add a subscript. One more rule: there is no such thing as a D_{nv} point group, so if you have this, call it instead a D_{nd} molecule. The official definition is that if the *v* planes bisect C_2 axes, then these are dihedral reflections, and are denoted *d*. The only way to really understand point groups is to get lots of practice. Look at the molecules below and think about what point group you would assign.

Benzene: C_6H_6

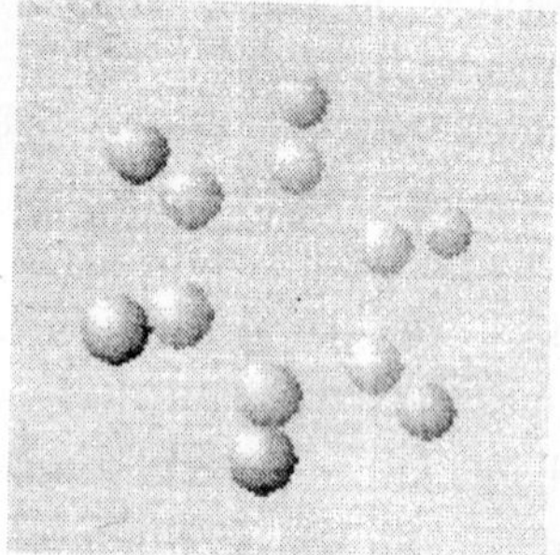

Below are some alternative representations of benzene. The last one shows the p orbitals making up benzene's conjugated system. (A conjugated system is a system in which there are alternating double and single bonds. Recall that a cumulated system, like allene, has adjacent double bonds.)

  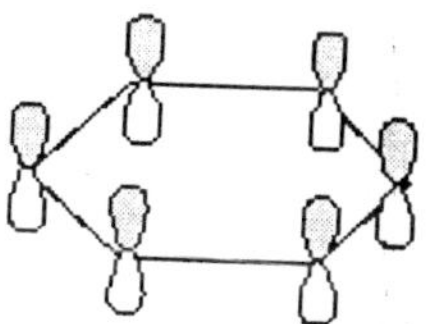

Cyclohexane: C_6H_{12}

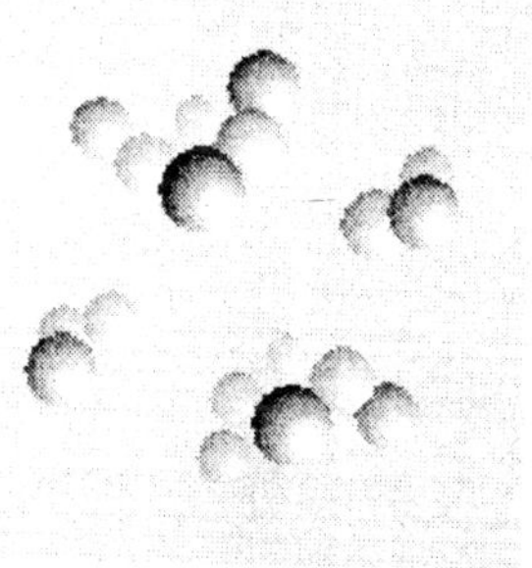

Below are some alternative representations of cyclohexane. Notice that unlike benzene, cyclohexane is not flat, but adopts a puckered shape that chemists refer to as a "chair." Below are two conformations of benzene. We will not go into how they differ and how they interconvert, but trust me, it's very cool.

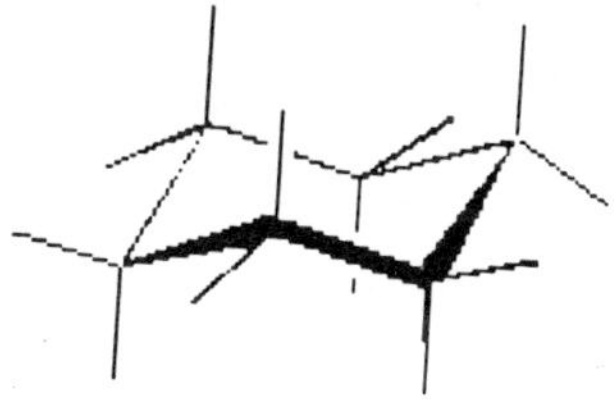 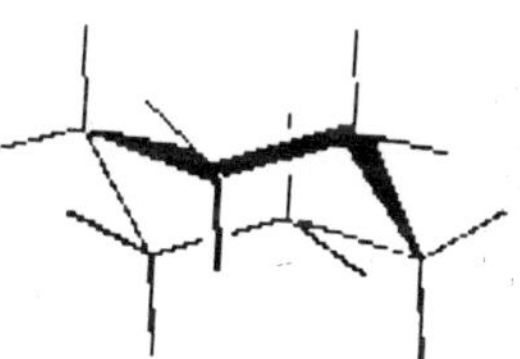

Water: H_2O

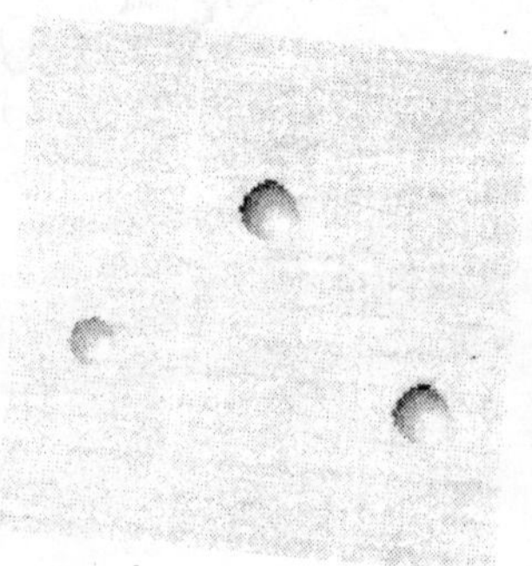

Ethane: C_2H_6

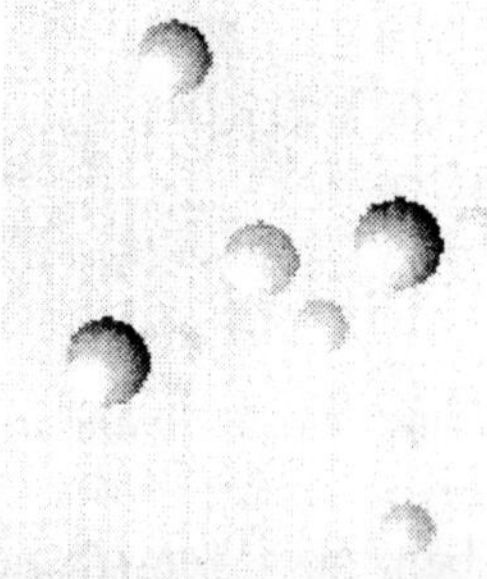

Note that ethane has two conformations with appreciable symmetry: one in which the hydrogen atoms are eclipsed and another in which they are staggered. In the figure above they are staggered. Use the mouse to get the picture to look like the Newman projection for staggered ethane that we looked at earlier. The conformations differ only by rotation about the carbon—carbon bond, which happens easily at room temperature. Just for fun, try to imagine what the symmetry elements would be for both conformations.

Buckminsterfullerine: C_{60}

If we can find all the symmetry elements of buckminsterfullerine. Buckminsterfullerene is an icosahedron. Following is a summary of the chemical applications of group theory that we will explore in this course. Note that these are only a few of the possibilities; group theory can be extremely powerful when it is applied to chemical systems. In fact, there is a whole book titled Chemical Applications of Group Theory that you will encounter if you continue on in physical and inorganic chemistry. As you delve deeper into science and math, you will gain an increasingly deep appreciation for the power and beauty of symmetry in the universe. But before we start waxing philosophical, here is that list:

— Only molecules belonging to the C_n, C_{nv}, and C_s point groups are polar, which means that they have a dipole moment.

You can usually tell just by looking at a molecule whether it has a dipole moment, but it's always nice to be able to back it up mathematically. And sometimes it's not so obvious from inspection. For example, tetrahedral molecules such as carbon tetrachloride (CCl_4) are non-polar despite the four very electronegative chlorines pointing off in different directions. If you were to calculate the dipole moment vectors, they would cancel each other out. But if you do not want to do the calculation, it is helpful to just know from group theory that tetrahedral molecules are non-polar and leave it at that.

— If a molecule has an improper axis of rotation, then it is achiral. Recall that the most common improper rotation axes are S_1, which is the same as a sigma mirror plane, and S_2, which is equivalent to inversion.

— Probably the most important chemical application of group theory is its application to bonding. We will only touch the tip of the iceberg regarding group theory and bonding in this course. What you will need to know for Chem 32 is that two orbitals must have the same symmetry about the internuclear axis in order to bond. A sigma bond, for example, is defined to be made up of two orbitals with C-infinity symmetry about the internuclear axis. A pi bond is made up of two orbitals with C_2 antisymmetry about the internuclear axis. C_2 antisymmetry means that a C_2 rotation would invert the sign of the wavefunction (i.e. flip the shading) in the orbitals making up the pi bond

Raman Spectroscopy

We are already aware that photons interact with molecules to induce transitions between energy states. In the discussion of Raman spectroscopy, we use language from particle theory and we say that a photon is scattered by the molecular system. Most photons are elastically scattered, a process which is called Rayleigh scattering. In Rayleigh scattering, the emitted photon has the same wavelength as the absorbing photon. Raman Spectroscopy is based on the Raman effect, which is the inelastic scattering of photons by molecules. The effect was discovered by the Indian physicist, C. V. Raman in 1928. The Raman effect comprises a very small fraction, about 1 in 10^7, of the incident photons. In Raman scattering, the energies of the incident and scattered photons are different.

The energy of the scattered radiation is less than the incident radiation for the Stokes line and the energy of the scattered radiation is more than the incident radiation for the anti-Stokes line. The energy increase or decrease from the excitation is related to the vibrational energy spacing in the ground electronic state of the molecule and therefore the wavenumber of the Stokes and anti-Stokes lines are a direct measure of the vibrational energies of the molecule. A schematic Raman spectrum may appear as:

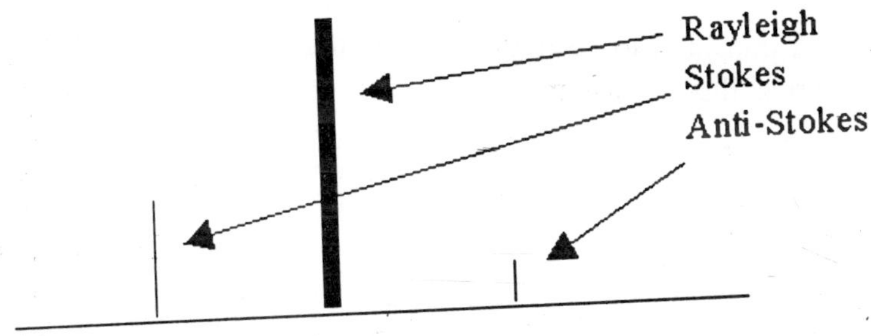

ω

In the example spectrum, notice that the Stokes and anti-Stokes lines are equally displaced from the Rayleigh line. This occurs because in either case one vibrational quantum of energy is gained or lost. Also, note that the anti-Stokes line is much less intense than the Stokes line. This occurs because only molecules that are vibrationally excited prior to irradiation can give rise to the anti-Stokes line. Hence, in Raman spectroscopy, only the more intense Stokes line is normally measured. Infrared (IR) and Raman spectroscopy both measure the vibrational energies of molecules but these method rely only different selection rules. Recall that for a vibrational motion to be IR active, the dipole moment of the molecule must change. Therefore, the symmetric stretch in carbon dioxide is not IR active because there is not change in the dipole moment. The asymmetric stretch is IR active due to a change in dipole moment.

no change in dipole

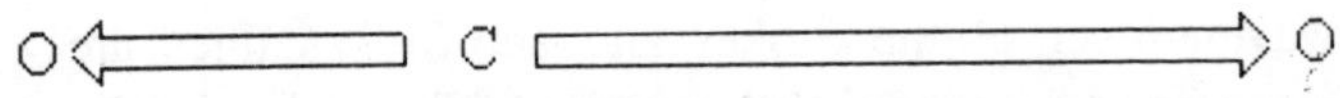

change in dipole

For a transition to be Raman active, there must be a change in polarizability of the molecule on next page. Notice that the symmetric stretch in carbon dioxide is Raman active because the polarizability of the molecule changes. You can see when you compare the ellipsoid at the equilibrium bond length to the ellipsoid for the extended and compressed symmetric motions. For a vibration to be Raman active, the polarizability of the molecule must change with the vibrational motion. Thus, Raman spectroscopy complements IR spectroscopy. Experimentally, we only observe the Stokes shift in a Raman spectrum. Recall that the Stokes lines will be at smaller wavenumbers (or higher wavelengths) than the exciting light. Since the Raman scattering is not very efficient, we need a high power excitation source such as a laser. Also, since we are interested in the energy (wavenumber) difference between the excitation and the Stokes lines, the excitation source should be monochromatic. This is another property of many laser systems.

Vibrational motion

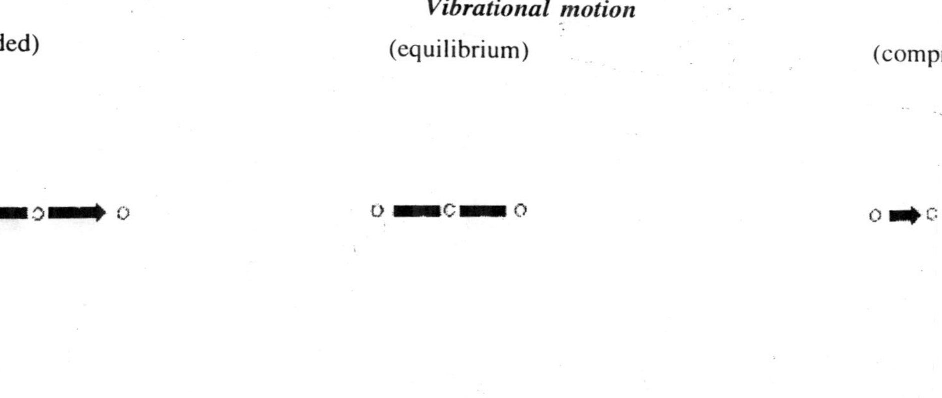

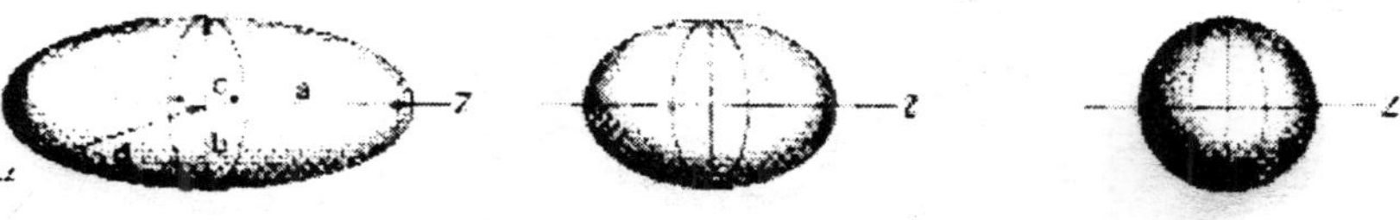

polarizability ellipsoid

Vibrational spectroscopy of molecules can be relatively complicated. Quantum mechanics requires that only certain well-defined frequencies and atomic displacements are allowed. These are known as the normal modes of vibration of the molecule. A linear molecule with N atoms has 3N - 5 normal modes, and a non-linear molecule has 3N - 6 normal modes of vibration. There are several types of motion that contribute to the normal modes. Some examples are:

— stretching motion between two bonded atoms;

— bending motion between three atoms connected by two bonds;

— out-of-plan deformation modes that change an otherwise planar structure into a non-planar one.

Infrared spectroscopy allows one to characterize vibrations in molecules by measuring the absorption of light of certain energies that correspond to the vibrational excitation of the molecule from $v = 0 \rightarrow v = 1$ (or higher) states. As indicated above, not all of the normal modes of vibration can be excited by infrared radiation. There are selection rules that govern the ability of a molecule to be detected by infrared spectroscopy.

The Raman effect was originally observed in 1928. It is due to the interaction of the electromagnetic field of the incident radiation, E_i, with a molecule. The electric field may induce an electric dipole in the molecule, given by

$$p = \alpha E_i \tag{1}$$

where is referred to as the polarizability of the molecule and p is the induced dipole. The electric field due to the incident radiation is a time-varying quantity of the form

$$E_i = E_0 \cos(2\pi v_i t) \tag{2}$$

For a vibrating molecule, the polarizability is also a time-varying term that depends on the vibrational frequency of the molecule, ν_{vib}

$$\alpha = \alpha_0 + \alpha_{vib} \cos(2\pi\nu_{vib}t) \tag{3}$$

Multiplication of these two time-varying terms, E_i and α, gives rise to a cross product term of the form:

$$\frac{\alpha_{vib}E_o}{2}[\cos 2\pi t(\nu_i + \nu_{vib}) + \cos 2\pi t(\nu_i - \nu_{vib})] \tag{4}$$

This cross term in the induced dipole represents light that can be scattered at both higher and lower energy than the Rayleigh (elastic) scattering of the incident radiation. The incremental difference from the frequency of the incident radiation, ν_i, are by the vibrational frequencies of the molecule, ν_{vib}. These lines are referred to as the "anti-Stokes" and "Stokes" lines, respectively. The ratio of the intensity of the Raman anti-Stokes and Stokes lines is predicted to be

$$\frac{I_A}{I_S} = \left(\frac{\nu_i + \nu_{vib}}{\nu_i - \nu_{vib}}\right)^4 e^{\left(\frac{-h\nu_{vib}}{kT}\right)} \tag{5}$$

The Boltzmann exponential factor is the dominant term in equation (5), which makes the anti-Stokes features of the spectra much weaker than the corresponding Stokes lines.

Infrared spectroscopy and Raman spectroscopy are complementary techniques, because the selection rules are different. For example, homonuclear diatomic molecules do not have an infrared absorption spectrum, because they have no dipole moment, but do have a Raman spectrum, because stretching and contraction of the bond changes the interactions between electrons and nuclei, thereby changing the molecular polarizability. For highly symmetric polyatomic molecules possessing a center of inversion (such as benzene) it is observed that bands that are active in the IR spectrum are not active in the Raman spectrum (and vice-versa). In molecules with little or no symmetry, modes are likely to be active in both infrared and Raman spectroscopy.

Point Groups. Molecules can be classified according to symmetry elements or operations that leave at least one common point unchanged. This classification gives rise to the point group representation for the molecule. Very useful information about the point group is contained in character tables. In this experiment we will study three different molecules: $CHCl_3$, CH_2Cl_2, and CH_3Cl. Both $CHCl_3$ and CH_3Cl are represented by the C_{3v} point group and CH_2Cl_2 is represented by the C_{2v} point group.

Infrared Transitions

For a fundamental transition to occur by absorption of infrared radiation the transition moment integral must be nonzero. The transition moment integrals are of the form:

$$\int \Psi_v^0 x \Psi_v^f d\tau$$

$$\int \Psi_v^0 y \Psi_v^f d\tau$$

$$\int \Psi_v^0 z \Psi_v^f d\tau$$

where Ψ_v^0 is the wave function for the initial state involved in the transition (the ground state), and Ψf is the wave function for the final state involved in the transition (the excited state). The *x*, *y* and *z* involved in the integrals refers to the Cartesian components of the oscillating electric vector of the radiation. If any of these three integrals is nonzero, then the transition moment integral is nonzero and the transition is allowed. We will use symmetry considerations to determine whether the transition moment integral is zero or nonzero, and hence whether the transitions is allowed or forbidden. The ground state wave function, Ψ_v^0, belongs to the totally symmetric representation of the point group, A_1 for the C_{2v} and C_{3v} point groups shown above. The symmetry representation for the excited state wave function, Ψ_v^f, depends on the symmetry of the normal mode vibration to be excited. If the product of the three terms in the transition moment integrals above are not the totally symmetric representation, then the integral will be zero. This leads to a very simple rule for the activity of fundamentals in infrared absorption:

A fundamental transition will be infrared active (that is, give rise to an absorption band) if the normal mode involved belongs to the same symmetry representation as any one or several of the Cartesian coordinates.

For the C_{2v} point group, this means that if the symmetry representation of the normal mode is A_1, B_1 or B_2, it will be infrared allowed. Only normal modes with the A_2 symmetry representation would be infrared forbidden. For the C_{3v} point group, this means that if the symmetry representation of the normal mode is A_1 or E, it will be infrared allowed. Only normal modes with the A_2 symmetry representation would be infrared forbidden.

Raman Transitions

For a fundamental transition to occur by Raman scattering of radiation the transition moment integral must be nonzero. The transition moment integrals are of the form:

$$\int \Psi_v^0 \alpha \Psi_v^f d\tau$$

where α represents the polarizability of the molecule. The symmetry representations for the polarizability is the same as that of quadratic terms involving the Cartesian coordinates, x^2,

y^2, z^2, xy, yz, and xz. These α's are components of the polarizability tensor and the requirement that the above integrals be nonzero means that there must be a change in polarizability of the molecule when the transition occurs. This leads to a very simple rule for the Raman activity of fundamentals:

A fundamental transition will be Raman active (that is, give rise to a Raman shift) if the normal mode involved belongs to the same symmetry representation as any one or more of the Cartesian components of the polarizability tensor of the molecule.

For the C_{2v} point group, this means that all of the symmetry representations of the normal mode: A_1, A_2, B_1 and B_2, will be Raman allowed. No normal modes for molecules in the C_{2v} point group will be Raman forbidden. For the C_{3v} point group, this means that if the symmetry representation of the normal mode is A_1 or E, it will be Raman allowed. Only normal modes with the A_2 symmetry representation would be Raman forbidden.

Normal Modes of Vibration

The normal modes of vibration for CH_3Cl and $CHCl_3$ (both molecules have the same symmetry) are shown in the figure on next page. The figure is from Herzberg, Infrared and Raman Spectra of Polyatomic Molecules, 1945. This shows that the first three vibrations for these C_{3v} molecules are of A_1 symmetry, and that the other three are doubly degenerate E symmetry vibrations.

In this experiment you will collect infrared and Raman spectra of liquid samples of the three chloromethanes: methyl chloride, methylene chloride and chloroform. The infrared spectra will be collected using a liquid cell and the FTIR spectrometer. Raman spectra will be collected using the argon ion laser - PTI spectrometer system show in the photograph. Liquid samples are placed in a fluorescence cell and in the sample compartment for the Raman experiments. The excitation of the Raman spectra uses the 488 nm line from the argon ion laser.

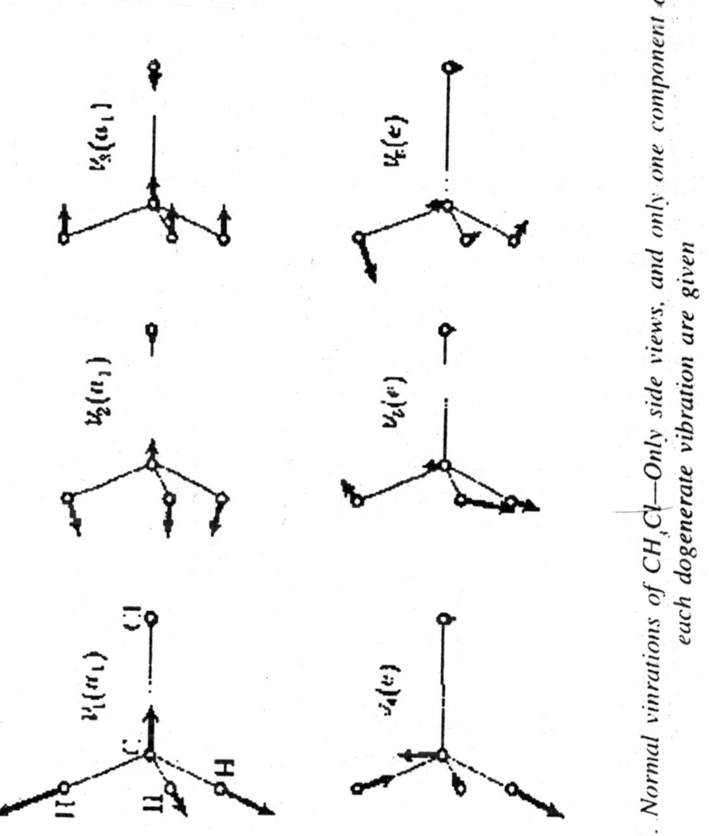

Fig. 1. Normal vinrations of CH_3Cl*—Only side views, and only one component of each dogenerate vibration are given*

A computerized spectrometer control and data acquisition is used for this experiment. A photograph is shown of the computerized spectrometer control and data acquisition program. Data can be saved to a data file that can be easily imported into a spreadsheet program for further analysis. The data file consists of a column of wavelengths and a column of signal intensities. These signal intensities are an average of 100 reading of the photomultiplier (PMT) signal. Be sure to record the photomultiplier voltage, and the amplifier and time constant settings. We will need to analyze the Raman data to construct a Raman spectra for the Stokes portion of the spectra. Determine the locations of the fundamental vibrations observed in the Raman spectra. Determine the locations of the fundamental vibrations observed in the infrared spectra. It will be necessary for you to use the literature to help assign the peaks to the particular vibrations in the molecule. Compare the infrared, Raman and literature values of the fundamental frequencies for the transitions for each of the molecules. Are any of the transitions that are infrared or Raman forbidden observed? Discuss the similarities and differences in the data obtained in the infrared and Raman spectra.

Normal Modes of Vibration

The complex vibrations of a molecule are the superposition of relatively simple vibrations called the normal modes of vibration. Each normal mode of vibration has a fixed frequency. It is easy to calculate the expected number of normal modes for a molecule made up of N atoms.

Linear molecule of N atoms: # normal modes = 3N - 5

Nonlinear molecule of N atoms: # normal modes = 3N - 6

The symmetries of the normal modes can be classified by group theory. As an example, water has a symmetrical bent structure of C_{2v} symmetry. It has three atoms and three normal modes of vibration (3*3 - 6 = 3).

Vibrational Spectroscopy

There are two types of spectroscopy that involve vibrational transitions. You should be very familiar with one of these from your Organic Chemistry course — infrared spectroscopy. During infrared spectroscopy experiments we observe transitions between vibrational energy levels of a molecule induced by the absorption of infrared (IR) radiation. The second type of vibrational spectroscopy is Raman spectroscopy. In Raman spectroscopy, vibrational transitions occur during the scattering of light by molecules. At room temperature almost all molecules are in their lowest vibrational energy levels with quantum number $n = 0$. For each normal mode, the most probable vibrational transition is from this level to the next highest level (n = 0 -> 1). The strong IR or Raman bands resulting from these transitiions are called fundamental bands. Other transitions to higher excited states (n = 0 -> 2, for instance) result in overtone bands. Overtone bands are much weaker than fundamental bands. Not all fundamental vibrational transitions can be studied by both IR and Raman specroscopy because they have different selection rules. Selection rules tell us if a transition is allowed or forbidden. An allowed transition has a high probability of occurring and will result in a strong band. Conversely a forbidden transition's probability is so low that the transition will not be observed. If a normal mode has an allowed IR transition, we say that it is IR active. Similarly if a normal mode has an allowed Raman transition, we say that it is Raman active.

If you know the point group of the molecule and the symmetry labels for the normal modes, then group theory makes it easy to predict which normal modes will be IR and/or Raman active. Look at the character table for the point group of the molecule.

If the symmetry label of a normal mode corresponds to x, y, or z, then the fundamental transition for this normal mode will be IR active. If the symmetry label of a normal mode corresponds to products of x, y, or z (such as x^2 or yz) then the fundamental transition for this normal mode will be Raman active.

Polarized and Depolarized Raman Bands

The assignment of Raman lines may be aided by measuring their intensity with a polarizing filter, first parallel and the perpendicular to the polarization of the incident radiation. If the polarization of the scattered beam is the same as that of the incident beam (intense only in the parallel direction), then the Raman line is said to be polarized. If the scattered light is intense in both the parallel and perpendicular direction, then the Raman line is depolarized.

The Geometry of the Sulfur Dioxide Molecule

Consider three possible geometries for the SO_2 molecule, linear and bent, shown at the right with their point groups.

O
H H
H H
A_1
B_2
Bent (C_{2v})

The symmetry types for the normal modes of the three structures are shown here. For the C_s form 3 A' means that there are three different normal modes, all having the same symmetry (A').

Similarly, for the C_{2v} form two of the three normal modes have the same symmetry (A_1). These modes are not identical and do not have the same energy—they just happen to have the same symmetry.

Structure	*Normal Modes*
C_s	$3A'$
C_{2v}	$2A_1$, B_2
$D_{\infty h}$	Σ_g^+, Σ_u^+, Π_u

Now we need to look at the character tables to see which normal modes one would expect to be observed in the IR and Raman for each structure. The character tables for the three point groups are shown below.

Cs structure: 3 normal modes, all having A' symmetry

	C_s	E	σh		
IR Raman	A'	1	1	x_1 y_1 R_z	x^2, y^2, z^2 xy
	A"	1	-1	z, R_x, R_y	yz, xz

The C_s structure should have 3 IR active fundamental transitions. These three fundamental transitions also should be Raman active. We would expect to observe three strong peaks in the IR and three strong peaks in the Raman at the same frequency as in the IR. All of the Raman lines would be polarized because they are totally symmetric (A' symmetry).

C_{2v} structure: 3 normal modes, two with A_1 symmetry, one with B_2

	C_{2v}	E	C_2	$\sigma_v(xz)$	$\sigma' v(yz)$		
IR Raman	A_1	1	1	1	1	z	z^2, y^2, x^2
	A_2	1	1	-1	-1	Rz	xy
	B_1	1	-1	1	-1	x, Ry	xz
IR Raman	B_2	1	-1	-1	1	y, Rx	yz

The C_{2v} structure should have 3 IR active fundamental transitions.

The three fundamental transitions also should be Raman active.We would expect to observe three strong peaks in the IR and three strong peaks in the Raman at the same frequency as in the IR.

Two of the Raman lines are totally symmetric (A_1 symmetry) and would be polarized. One Raman line would be depolarized.

$D_{\infty h}$ structre: 3 normal modes with Σ^+_g, Σ^+_u, Π_u symmetries

	$D_{\infty h}$	E	$2C^{F}_{\infty}$			
Raman	Σ^+_g	1	1	....		x^2+y^2, z^2
	Σ^-_g	1	1	....	R_z	
	Π_g	2	2 cos Φ	....	(R_x, R_y)	(xz, yz)
	Δ^g	2	2 cos2Φ	....		$(x^2- y^2, xy)$
IR	Σ^+_u	1	1	....	z	
	Σ^-_u	1	1	....		
IR	Π_u	1	1	....	(x, y)	
	Δ_u	2	2 $\cos^2\Phi$	....		

The $D_{\infty h}$ structure should have two IR active fundamental transitions. It will have one *Raman* active fundamental transition at a different frequency than either of the IR peaks. The *Raman* line will be polarized.

The experimental infrared and Raman bands of liquid and gaseous sulfur dioxide have been reported in a book by Herzberg. Only the strong bands corresponding to fundamental transitions are shown below. The polarized Raman bands are in red.

Fundamental	2	1	3
IR (cm^{-1})	519	1151	1336
Raman (cm^{-1})	524	1151	1336

The existence of three experimental bands in the IR and Raman corresponding to fundamental transitions weighs strongly against the symmetrical linear ($D_{\infty h}$) structure. We usually do not expect more strong bands to exist than are predicted by symmetry. Group theory predicts that both bent structures would have three fundamental transitions that are active in both the IR and Raman. However all three of the Raman lines would be polarized if the structure were unsymmetrical (C_s symmetry). The fact that one Raman line is depolarized indicates that the structure must be bent and symmetrical (C_{2v} symmetry).

7

Group Theory and Quantum Mechanics

Functions as Basis for the Representation of a Group

So far we have considered the action of symmetry operations on the coordinates of points. In analogy we can define the action of the symmetry operation $\hat{R}$ on a function f:

$$\hat{R} f \vec{r} = f(\hat{R}^{-1} \vec{r})$$

Now we can use a set of n functions f_1, f_1,...,f_n. The set forms a basis for a representation of the group if the following holds:

$$\hat{R}\begin{pmatrix} f_1 \\ f_2 \\ \vdots \end{pmatrix} = \begin{pmatrix} f_1' \\ f_2' \\ \vdots \end{pmatrix} = \Gamma(R)\begin{pmatrix} f_1 \\ f_2 \\ \vdots \end{pmatrix}$$

$\Gamma(R)$ is a representation of the symmetry group in the n-dimensional basis f_1, f_1,...,f_n. We say:

f_1, f_1,...,f_n span a basis for a n-dimensional representation of the group.

Wave Functions as Basis for Irreducible Representations

Zerested in thc symmetry properties of a wave function 'I', which is solution of the Schrödinger equation

$$\hat{H}\Psi = E\Psi$$

We consider a symmetry operation $\hat{R}$ of the system. As the symmetry operation transform the system into an physically equivalent state, we expect that Hamilton operator $\hat{H}$ should be invariant with respect to the transformation into a new basis:

$$\hat{H} = \hat{R}^{-1}\hat{H}\hat{R} \text{ or } \hat{R}\hat{H} = \hat{H}\hat{R}$$

i.e. $\hat{H}$ and $\hat{R}$ commute.

This has important implications for the eigenfunctions of $\hat{H}$:

— *Case 1: non-degenerate eigenfinctions:*

Schrödinger equation: $\hat{H}\Psi = E_i\Psi_{ii}$

symmetry operation: $\hat{R}\hat{H}\Psi_i = \hat{R}\hat{H}_i\Psi_i$

$$\hat{H}(\hat{R}\Psi_i) = E_i(\hat{R}\Psi_i)$$

$\hat{R}\Psi_i$ is eigenfunction of $\hat{H}$ with eigenvalue E_i. As the system is non-degenerate (by definition), it follows: $\hat{H}\Psi = \pm\Psi_i$ i.e. non-degenerate eigenfunctions of $\hat{H}$ belong to a 1-dimensional irrep of the group.

— *Case 2: Degenerate Eigenfunction:*

Schrödinger equation: $\hat{H}\Psi_{il} = E_i\Psi_{il} \forall \Psi_{i,l,k}$

symmetry operation: $\hat{H}(\hat{R}_{\Psi il}) = E_i(\hat{R}\Psi_{il})$

$\hat{R}\Psi_i$ is eigenfunction of $\hat{H}$ with eigenvalue E_i. As the system is k-fold degenerate, $\hat{R}\Psi_i$ can be a linear combination of the eigenfunctions

$$\hat{R}\Psi_{i,l,k} \quad \hat{R}\Psi_{il} = \sum_{j=i}^{k} r_{ij}\Psi_j$$

i.e. a set of k degenerate eigenfunctions of $\hat{H}$ belongs to a k-dimensional irrep of the group (note: (1) any linear combination of eigenfunctions of degenerate levels is eigenfunction of the system as well; (2) if the representation was reducible, the functions would not have to be transformed into each other. In this case the eigenvalues would not have to be identical as well).

Direct Product of Functions

We assume two sets of functions

$$X_1, X_2, \ldots, X_m \text{ and } Y_1, Y_2, \ldots, Y_n,$$

which form a basis for a representation of the group (for example two sets of wavefunctions). What is the symmetry of the product functions X_iY_j (extremely important question, a product function occur often in quantum mechanics)?

We consider the application of symmetry operation $\hat{R}$:

$$\hat{R}X_i = \sum_{j=i}^{m} x_{ij} X_j \text{ and } \hat{R}Y_i = \sum_{j=i}^{n} y_{ij} Y_j$$

Application to product:

$$\hat{R}X_iY_k = \sum_{j=1}^{m} x_{ij} X_j \hat{R}Y_k = \sum_{j=1}^{m}\sum_{l=1}^{n} \underbrace{x_{ij}y_{lk}}_{z_{ji,kl}} X_jY_l \,.$$

$z_{ij,kj}$ is a matrix of dimension (nm) × (nm).

The functions X_iY_j with i = 1, 2,...n; j = 1,2,...m are called the direct product of X_i and Y_j (X ⊗ Y). They span a basis for a representation of the group of dimension mxn.

For the characters of the direct product we obtain:

$$\chi_z(R) = \sum_{j,l} z_{jl,jl} = \sum_j \sum_l x_{jj} y_{ll} = \chi_x(R)\chi_y(R)$$

i.e. the characters of the representation spanned by the direct product of two sets of functions are the products of the characters of the original representations.

Nonzero Matrix Elements

Why are direct products important? In quantum mechanics we are often interested in matrix elements of an operator, e.g. if we would like to determine the expectations value of an observable. These matrix elements are of the type, i.e. they contain direct products of functions (and operators):

$$\int f_A{}^* f_B d\tau \quad \text{or} \quad \langle f_A | f_B \rangle$$

$$\int f_A{}^* \hat{O} f_A d\tau \quad \text{or} \quad \left\langle f_A \middle| \hat{O} \middle| f_B \right\rangle.$$

Such matrix elements can only be nonzero, the function over which we integrate is completely symmetric or contains a completely symmetric part.

The representation of a direct product contains the totally symmetric representation only if the representations of the product functions are of identical symmetry (or at least contain a part with identical symmetry).

Proof: We consider two sets of functions A⊗Ä. The characters of the direct products are χ_{AB}. For an arbitrary irrep of the group i we obtain

$$a_i = \frac{1}{h}\sum_R \chi_i(R)\chi_{AB}(R)$$

for the number of irreps of symmetry type i, contained in A⊗Ä. For the totally symmetric irrep:

$$a_i = \frac{1}{h}\sum_R \chi_{AB}(R) = \frac{1}{h}\sum_R \chi_A(R)\chi_B(R) = \delta_{AB}$$

Consequently, A and B must be of identical symmetry in order to contain the totally symmetric representation.

Examples for Matrix Elements

Hamilton operator: $\langle A|\hat{H}|B\rangle$

As discussed before, $\hat{H}$ is totally symmetric with respect to the symmetry operations of the system, i.e. $\Gamma_{\hat{H}} = A_i$. Therefore:

$$\Gamma_{\langle f_A|\hat{H}|f_B\rangle} = \Gamma_i \otimes \Gamma_{\hat{H}} \otimes \Gamma_f = \Gamma_i \otimes \Gamma_f$$

Hamilton matrix elements of wave function of different symmetry are zero.

Dipole operator: $\langle A|\hat{\mu}|B\rangle$

Electronic dipole operator:

$$\hat{\mu} = \underbrace{\sum_i q_i x_i}_{\mu_x} + \underbrace{\sum_i q_i x_i}_{\mu_y} + \underbrace{\sum_i q_i x_i}_{\mu_z}$$

(q_i, $\vec{r}_i$ charge, position of particle i)

The parts of the dipole operator have the symmetry of the functions x, y, z.

Case A: Permanent dipole moment $\langle A|\hat{\mu}|A\rangle$

$$\Gamma_{\hat{\mu}x} = \Gamma_A \otimes \Gamma_A \otimes \Gamma_x$$
$$\Gamma_{\hat{\mu}y} = \Gamma_A \otimes \Gamma_A \otimes \Gamma_y$$
$$\Gamma_{\hat{\mu}z} = \underbrace{\Gamma_A \otimes \Gamma_A}_{\Gamma_A^2} \otimes \Gamma_z$$

A permanent dipole moment exists if the symmetry of x, y or z is contained in Γ_A^2. Note: Γ_A^2 always contains the totally symmetric representation.

Case B: Transition dipole moment $\langle f|\hat{\mu}|i\rangle$

A nonzero transition dipole moment requires that at least one of the components

$$\Gamma_{\hat{\mu}x} = \Gamma_i \otimes \Gamma_f \otimes \Gamma_x$$
$$\Gamma_{\hat{\mu}y} = \Gamma_i \otimes \Gamma_f \otimes \Gamma_y$$
$$\Gamma_{\hat{\mu}z} = \Gamma_i \otimes \Gamma_f \otimes \Gamma_z$$

contains the totally symmetric representation (IR, VIS, UV spectroscopy).

Projection Operators

The last section demonstrates that it is useful to construct wavefunctions, which represent a basis for the irreps of the symmetry group. If this is the case, we can easily decide which integrals are zero and which can be nonzero. How are these functions constructed in a general fashion?

Example:

We consider a set of atomic wave functions, i.e. the three 1s function located at the H atoms of a NH_3 molecule:

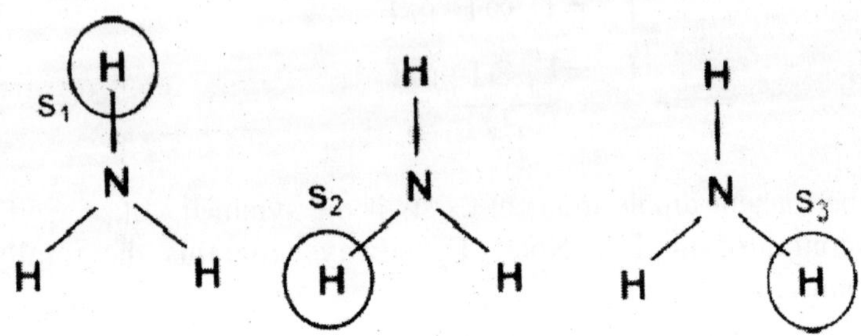

(1) These functions span a basis for a three dimensional representation of the group C_3V (i.e. the symmetry operations of the group transform the wave functions into linear combinations of each other).

(2) The representation is reducible (as there are only 1 and 2 dimensional representations in C_{3V}).

We have to perform a transformation to a new basis, which transforms the representation into a ret of irreps of C_{3V} (i.e. blockdiagonalizes the transformation matrices). This new symmetry adapted basis is constructed applying a set of projection operators:

We consider a set of l_i functions ϕ_i', ϕ_2', ..., ϕ_{li}', which correspond to the i-th irrep of the group which has dimension l_i. Now we apply an operation of the group R to one function ϕ'_i :

$$\hat{R}\, \text{fi}' = \sum_{x} \Gamma^{i}_{st}(\hat{R})\varphi^{i}_{s} \text{ (definition of the representation).}$$

By left-multiplication with / summation over operations of the group $\sum_{R} \Gamma^{j}_{s't'}(\hat{R})^{*}$

$$\sum_R \Gamma^j_{s't'}(\hat{R})^* \hat{R}\varphi^i_t = \sum_R \sum_s \Gamma^j_{s't'}(\hat{R})^* \Gamma^i_{st}(\hat{R})\varphi^i_s$$

$$= \sum_s \varphi^i_s \underbrace{\sum_R \Gamma^i_{st}(\hat{R})\Gamma^j_{s't'}(\hat{R})^*}_{\substack{\text{GOT (section 3.12)} \\ \frac{h}{\sqrt{l_i l_j}}\delta_{ij}\delta_{ss'}\delta_{tt'}}}$$

$$\Rightarrow \underbrace{\frac{l_j}{h}\sum_R \Gamma^j_{s't'}(\hat{R})^* \hat{R}}_{\hat{P}^j_{s't'}} \varphi^i_t = \delta_{ij}\delta_{tt'}\varphi^i_{s'}$$

Interpretation: $\hat{P}^j_{s't'}$ acts on some function φ^i_t. Only if this function contains a component which belongs to the irrep j and within this irrep exactly to the function t', the operation will yield the s'-th function if irrep j.

Most important case:

The operator acts on a function φ^i_t. Only if this function contains the t'-th function of irrep j the operation yields φ^i_t (i.e. the t'-th function of irrep j). In other words: the operator "projects" the component of desired symmetry out of the starting function.

Consequently, we can generate the desired set of symmetry adapted functions by applying all projection operators to the full set of function spanning the representation. The drawback: we need the full representation matrices, but all we have are the characters. Therefore we construct a simplified projection operator:

By summing over the functions contained in irrep j:

$$\hat{P}_j = \sum_{l'} \hat{P}^j_{l'l'}$$

$$= \frac{l_j}{h} \sum_{l'} \sum_{R} \Gamma^j_{l'l'}\left(\hat{R}\right)^* \hat{R}$$

$$= \frac{l_j}{h} \sum_{R} \underbrace{\sum_{l'} \Gamma^j_{l'l'}\left(\hat{R}\right)^*}_{\chi^j\left(\hat{R}\right)} \hat{R}$$

$$\Rightarrow \boxed{\hat{P}^j = \frac{l_j}{h} \sum_{R} \chi^j\left(\hat{\mathrm{R}}\right) \hat{R}}$$

8

Recent Advances in Perturbation Theory

Multireference techniques can handle nondynamical correlation well. Once the state-specific nondynamical correlation is removed, the rest is primarily composed of dynamical pair correlation and individual pair correlation, and can be described even by second-order perturbation theory. This is the basic idea of the multireference-based perturbation theory (MRPT). Multireference Møller-Plesset (MRMP) and quasidegenerate perturbation theory (MC-QDPT) have been successfully applied to many chemical and spectroscopic problems, and this approach has established itself as an efficient method for treating nondynamical and dynamical correlation effects. MRMP can handle any state, regardless of charge, spin, or symmetry, with surprisingly high and consistent accuracy. However, MRMP has a sharp limit to the number of configurations of the reference complete active space (CAS) SCF wave function. To avoid this problem, many approaches have been proposed.

We have developed perturbation theory (PT) based on the quasicomplete active space (QCAS) SCF wave function. QCAS is defined as the product space of CAS spanned by the determinants or configuration state functions (CSFs). Although

QCAS works quite well, QCAS requires physically sound judgment and intuition in the choice of subspace. More flexible reference functions are required. A second-order PT starting with general multiconfiguration (MC) SCF wave functions has been developed. The general MCSCF functions are wave functions optimized in an active space spanned by an arbitrary set of Slater determinants or CSFs.

The approach can dramatically reduce the dimension of the reference function. Recently, a very efficient string product space (SPS) SCF/PT has been proposed, where the total space is defined as a product of α and β string spaces. This review will focus on our recent development in multireference-based perturbation theory.

Multireference Møller-Plesset Perturbation Method

Our basic problem is to find approximations to some low-lying solutions of the exact Schrödinger equation,

$$H\Psi = E\Psi \tag{1}$$

H is the Hamiltonian and it is decomposed into two parts, a zeroth-order Hamiltonian H_0 and a perturbation V,

$$H = H_0 + V \tag{2}$$

We assume that a complete set of orthonormal eigenfunctions and corresponding eigenvalues is available,

$$H_0\Psi_i^{(0)} = E_i^{(0)}\Psi_i^{(0)} \tag{3}$$

Then the state wave function Ψ_l is expanded in terms of basis functions $\Psi_k^{(0)}$ as

$$\Psi_l = \sum_k C_{lk}\,\Psi_k^{(0)} \tag{4}$$

Some of the basis functions define an active space P, and the remaining part of Hilbert space is called the orthogonal space $Q = 1 - P$. The active space is spanned by the basis functions that have a filled core and the remaining active electrons distributed over a set of active orbitals. The orthogonal complete space incorporates all other possible basis functions that are characterized by having at least one vacancy in a core orbital. The state wave function in an active space is written as

$$\Psi_k^{(0)} = \sum_k C_k \Phi_k, \tag{5}$$

where the sum runs over the active space basis functions $\{\Phi_i\}$ and C_k are the coefficients of just the active space basis functions. It is convenient to use intermediate normalization, i.e.

$$\langle \Psi_I^{(0)} | \Psi_I^{(0)} \rangle = \langle \Psi_I^{(0)} | \Psi_I \rangle = 1 . \tag{6}$$

We also assume $\Psi_I^{(0)}$ that is diagonal in P space,

$$\langle \Psi_I^{(0)} | H | \Psi_J^{(0)} \rangle = \delta_{IJ} (E_I^{(0)} + E_I^{(1)}) , \tag{7}$$

with

$$E_I^{(0)} = \langle \Psi_I^{(0)} | H_0 | \Psi_I^{(0)} \rangle , \tag{8}$$

$$E_I^{(0)} = \langle \Psi_I^{(0)} | V | \Psi_I^{(0)} \rangle . \tag{9}$$

The state-specific Rayleigh-Schrödinger PT based on the unperturbed eigenvalue equation

$$H_0 \Psi_I^{(0)} = E_I^{(0)} \Psi_I^{(0)} \tag{10}$$

leads to the first $E_I^{(k)}$ few as

$$E_I^{(2)} = \langle \Psi_I^{(0)} | VRV | \Psi_I^{(0)} \rangle , \tag{11}$$

$$E_I^{(3)} = \langle \Psi_I^{(0)} | VR(V - E_I^{(1)}) RV | \Psi_I^{(0)} \rangle , \tag{12}$$

$$\begin{aligned} E_I^{(4)} &= \langle \Psi_I^{(0)} | VR(V - E_I^{(1)}) R(V - E_I^{(1)}) RV | \Psi_I^{(0)} \rangle \\ &\quad - E_I^{(2)} [\langle \Psi_I^{(0)} | VR^2 V | \Psi_I^{(0)} \rangle + \langle \Psi_I^{(0)} | VRH_0 SH_0 RV | \Psi_I^{(0)} \rangle] \end{aligned} \tag{13}$$

R and S are the resolvent operators

$$R = Q/(E_I^{(0)} - H_0) , \tag{14}$$

$$S = P'/(E_I^{(0)} - H_0) , \tag{15}$$

where $P' = P - |\Psi_1^{(0)}\rangle \langle \Psi_1^{(0)}|$.

$E_1^{(0)}$ is given in terms of orbital energies as

$$E_I^{(0)} = \sum_k D_{kk} \varepsilon_k , \tag{16}$$

and the orbital energies are defined as

$$\varepsilon_i = \langle \varphi_i | F | \varphi_i \rangle \tag{17}$$

with

$$F_{ij} = h_{ij} + \sum_{kl} D_{kl}\left[(ij|kl) - \frac{1}{2}(ik|lj)\right], \tag{18}$$

where D_{ij} is the one-electron density matrix. The MCSCF orbitals are resolved to make the F_{ij} matrix as diagonal as possible. This zeroth-order Hamiltonian is closely analogous to the closed-shell Fock operator. The definition of an active space, the choices of active orbitals and the specification of the zeroth-order Hamiltonian completely determine the perturbation approximation.

When a CASSCF wave function is used as the reference, the zeroth plus first order energy $E_1^{(0)} + E_l^{(1)}$ is equal to the CASSCF energy. The lowest non-trivial order is therefore the second order. Let the reference function $|\Psi_\alpha^{(0)}$ be a CASSCF wave function,

$$|\alpha\rangle = \sum_{k} C_A|A\rangle. \tag{19}$$

The energy up to the second order is given by

$$E_\alpha^{(0-2)} = E_\alpha^{\mathrm{CAS}} + \sum_{I} \frac{\langle\alpha|V|I\rangle\langle I|V|\alpha\rangle}{E_\alpha^{(0)} - E_I^{(0)}} \tag{20}$$

where is the set of all singly and doubly excited configurations from the reference configurations in CAS. This is our multireference Møller-Plesset (MRMP) method. We have also proposed a multistate multireference perturbation theory, the quasidegenerate perturbation theory with MCSCF reference functions (MC-QDPT).

The Perturbation Theory Based on QCAS-SCF and MCSCF Wave Functions

CASSCF can handle the near-degeneracy problem in a balanced way and therefore can treat chemical reactions and excited states. Once the active space is chosen, the wave function is completely specified. It is size-consistent and the wave function is invariant to transformations among active orbitals. Although CASSCF does not include dynamical correlation, it provides a good starting point for such studies. However, CASSCF often generates too many configurations with the number of active orbitals and active electrons. To reduce the CAS dimension, we have proposed the quasi-complete active space (QCAS) SCF method. QCAS is an attempt to extend the method to widen the range of applications. QCAS is defined as a product of complete active spaces. Let us divide the active electrons and orbital sets into N subsets and fix the number of active electrons, m_i, and orbitals, n_i, in each subset,

$$m_{act} = \sum_{i}^{N} m_i, \quad n_{act} = \sum_{i}^{N} n_i \tag{21}$$

where m act and n act denote the number of active electrons and active orbitals, respectively. We define the QCAS as the product space of CAS spanned by the determinants or CSF

$$QCAS(\{m_i\},\{n_i\}) = CAS(m_1,n_1) \times \mathrm{CAS}(m_2,n_2) \times ... \times \mathrm{CAS}(m_N,n_N) \tag{22}$$

Each active space is defined by a fixed number of active electrons and active orbitals. Solution of the CI eigenvalue problem involves the σ_1 vector, which is given by

$$\sigma_1 = \sum_{k} H_{IJ} C_J$$

$$= \sum_J \left\{ \sum_{ij} h_{ij} \langle I|E_{ij}|J\rangle + \frac{1}{2} \sum_{ijkl} (ij|kl) \langle I|E_{ij}E_{kl} - \delta_{jk}E_{il}|J\rangle \right\} C_J \tag{23}$$

Here, I and E_{ij} are the CI basis functions and the group generators. We adopt Slater determinants rather than CSF and split a determinant into α and β strings of each group

$$|I\rangle = |I_\alpha^1 \cdots I_\alpha^N; I_\beta^1 \cdots I_\beta^N\rangle = |I_\alpha^1\rangle \cdots |I_\alpha^N\rangle |I_\beta^1\rangle \cdots |I_\beta^N\rangle \tag{24}$$

Then we can decompose the one-body a and b coupling constants into the coupling constants for the strings of the groups as

$$\langle I|E_{ij}|J\rangle = \langle I|E_{ij}^\alpha|J\rangle + \langle I|E_{ij}^\beta|J\rangle$$

$$= \begin{cases} \langle I_\alpha^G|E_{ij}^\alpha|J_\alpha^G\rangle \prod_{H\neq G} \delta_{I_\alpha^H J_\alpha^H} \prod_H \delta_{I_\beta^H J_\beta^H} + \langle I_\beta^G|E_{ij}^\alpha|J_\beta^G\rangle \prod_H \delta_{I_\alpha^H J_\alpha^H} \prod_{H\neq G} \delta_{I_\beta^H J_\beta^H} & (i,j \in G) \\ 0 & (otherwise) \end{cases} \tag{25}$$

Thus, the σ-vector and the one- and two-particle density matrices are expressed by the coupling constants for the strings of the groups. The dimension of QCAS is much smaller than that of CAS constructed from the same set of electrons and orbitals. Let us consider CAS(16e, 16o), where 16 electrons are distributed among 16 orbitals. CAS(16e, 16o) is spanned by the 165 636 900 determinants (with M = 0). If we divide the active electrons and orbitals into five groups: (4e, 4o) + (4e, 4o) + (4e, 4o) + (2e, 2o) + (2e, 2o), the dimension of QCAS is reduced to 746 496. Using QCAS as a reference function in the perturbation theory, we may therefore extend active electrons and orbitals beyond the limit of CAS. QCASSCF/ PT works quite well. However, it is not always possible to select an appropriate QCAS, depending on the molecular systems of interest. QCAS requires physically sound judgment and intuition in the choice of subspace.

Therefore, we have developed a second-order QDPT using a general multiconfiguration (MC) SCF wave function as a reference function (hereafter, GMC-QDPT). The general configuration space (GCS) is defined by a space that is spanned by an arbitrary set of Slater determinants or CSFs. The orbitals are partitioned into three categories as in the ordinary MCSCF method: the core orbitals are doubly occupied and the virtual orbitals are unoccupied in all the determinants/CSFs, while the active orbitals may be occupied or unoccupied. The reference wave functions used in the perturbation calculations are determined by MCSCF as a variational space:

$$|\alpha\rangle = \sum_{A \in GCS} C_A(\alpha)|A\rangle . \tag{26}$$

The effective Hamiltonian up to the second order $H^{(0\text{-}2)}_{\text{eff}}$ of van Vleck perturbation theory with unitary normalization is given by

$$(H_{eff}^{(0-2)})_{AB} = H_{AB} + \frac{1}{2}[\langle\Phi_A^{(0)}|HR_BH|\Phi_B^{(0)}\rangle + \langle\Phi_A^{(0)}|HR_AH|\Phi_B^{(0)}\rangle] \tag{27}$$

with

$$R_A = \sum_{I \notin ref} |\Phi_I^{(0)}\rangle(E_A^{(0)} - E_I^{(0)})^{-1}\langle\Phi_I^{(0)}| , \tag{28}$$

where $\Phi_A^{(0)}$ ($\Phi_B^{(0)}$) and $\Phi_I^{(0)}$are reference wave functions and a function in the complement space (Q) of the reference space (P), respectively, and $E_B^{(0)}$and $E_I^{(0)}$ are zeroth-order energies of functions $\Phi_B^{(0)}$ and $\Phi_I^{(0)}$. Adopting (state-averaged) MCSCF wave functions a (b) as reference functions $\Phi_A^{(0)}$ and $\Phi_B^{(0)}$, which define the P space, Eq. (27) becomes

$$(K_{eff}^{(0-2)})_{\alpha\beta} = E_\alpha^{MC-SCF}\delta_{\alpha\beta} + \frac{1}{2}\sum_{I\notin GCS}\left\{\frac{\langle\alpha|H|I\rangle\langle I|H|\beta\rangle}{E_\beta^{(0)} - E_I^{(0)}} + (\alpha \leftrightarrow \beta)\right\}, \quad (29)$$

where I is a determinant/CSF outside the GCS. The notation ($\alpha \leftrightarrow \beta$) means interchange α with β from the first term in curly brackets. The complementary eigenfunctions of the MCSCF Hamiltonian and the determinants/CSFs generated by exciting electrons out of the determinants/CSFs in GCS are orthogonal to the reference functions and define the Q space. The functions in the space complementary to the P space, however, do not appear in Eq. (29), because the interaction between the complementary functions and the reference functions is zero. We define here the corresponding CAS (CCAS) as a CAS constructed from the same active electrons and orbitals, that is, the minimal CAS that includes the reference GCS. The summation over I in Eq. (29) may be divided into summations over the determinants/CSFs outside CCAS and over the determinants/CSFs outside the GCS but inside CCAS:

$$\sum_{I\notin GCS} = \sum_{I\notin CCAS} + \sum_{I\notin CCAS \wedge \notin GCS}, \quad (30)$$

then the former second-order term may be written as

$$(K_{eff}^{(2)})_{\alpha\beta} = \sum_{I \notin CCAS} \frac{\langle\alpha|H|I\rangle\langle I|H|\beta\rangle}{E_\beta^{(0)} - E_I^{(0)}}$$
$$+ \sum_{I \notin CCAS \wedge I \notin GCS} \frac{\langle\alpha|H|I\rangle\langle I|H|\beta\rangle}{E_\beta^{(0)} - E_I^{(0)}}. \tag{31}$$

The first term in Eq. (31) represents external excitations, and the second term represents internal excitations. The external term is calculated by the diagrammatic method and the internal term by the sum-over-states method. The external term may be further written as

$$(H_{ext}^{(2)})_{\alpha\beta} = \sum_{A,\, B \in GCS} C_A(\alpha) C_B(\beta) (H_{ext}^{(2)})_{AB} \tag{32}$$

with

$$(H_{ext}^{(2)})_{AB} = \sum_{I \notin CCAS} \frac{\langle A|H|I\rangle\langle I|H|B\rangle}{E_B^{(0)} - E_I^{(0)} + (E_\beta^{(0)} - E_B^{(0)})}, \tag{33}$$

where $(H_{ext}{}^{(2)}$ is the effective Hamiltonian in the determinant/CSFs. Because the second-order diagrams do not depend on the denominator, the second-order effective Hamiltonian, Eq. (33) (hence, also Eq. (32)) is expressed by the same diagrams as in the conventional QDPT. For internal terms, the diagrammatic approach may not be applied. Instead, matrix operations for the Hamiltonian matrix are used:

$$(H_{int}^{(2)})_{\alpha\beta} = \mathbf{v}^T(\alpha) \cdot \mathbf{w}(\beta) \tag{34}$$

with

$$\mathbf{v}_I(\alpha) = \sum_{A \in GCS} \langle I|H|A\rangle C_A(\alpha) \tag{35}$$

$$\mathbf{w}_I(\beta) = \sum_{B \in GCS} \langle I|H|B\rangle C_B(\beta)/(E_\beta^{(0)} - E_B^{(0)}) \tag{36}$$

The intermediate determinants/CSFs I are constructed by exciting one or two electrons from the reference determinants/CSFs within the active orbital space. In general, the number of I is not large, and thus they may be managed in computer memory. In the present implementation, we used Slater determinants rather than CSFs. Let $\{I_\alpha\}$ and $\{I_\beta\}$ be sets of α and β strings appearing in the reference configurations, respectively. The reference space is defined by the β string sets for each α string, $\{I_\beta[I_\alpha]\}$, and equivalently the string sets for each β string, $\{I_\alpha[I_\beta]\}$.

In the diagrammatic computation of the external terms, one-, two-, and three-body coupling coefficients are necessary. The one-body coupling coefficients are classified into two types,

$$\langle I_\alpha|E_{pq}^\alpha|J_\alpha\rangle\langle I_\beta|J_\beta\rangle \text{ and } \langle I_\alpha|J_\alpha\rangle\langle I_\beta|E_{pq}^\beta|J_\beta\rangle$$

the two-body coupling coefficients into three types,

$$\langle I_\alpha|E_{pq,rs}^\alpha|J_\alpha\rangle\langle I_\beta|J_\beta\rangle,\ \langle I_\alpha|J_\alpha\rangle\langle I_\beta|E_{pq,rs}^\beta|J_\beta\rangle, \text{ and}$$

$$\langle I_\alpha|E_{pq}^\alpha|J_\alpha\rangle\langle I_\beta|E_{rs}^\beta|J_\beta\rangle,$$

and the three-body coupling coefficients into four types,

$$\langle I_\alpha|E^{\alpha}_{pq,rs}|J_\alpha\rangle\langle I_\beta|E^{\beta}_{tu}|J_\beta\rangle\text{, and }\langle I_\alpha|E^{\alpha}_{pq}|J_\alpha\rangle\langle I_\beta|E^{\beta}_{rs,tu}|J_\beta\rangle$$

with $J_\alpha \in \{\hat{I}_\alpha\}$, $\{I_\beta\}$, and

$$E^{\alpha}_{pq,rs,\ldots} = a^{+}_{p\alpha}a^{+}_{r\alpha}\cdots a_{s\alpha}a_{q\alpha}, \tag{37}$$

$$E^{\beta}_{pq,rs,\ldots} = a^{+}_{p\beta}a^{+}_{r\beta}\cdots a_{s\beta}a_{q\beta}. \tag{38}$$

Because string J_σ (J_β) is determined by string $I_\alpha(I_\beta)$ and active orbital labels p and q, the one-body coupling coefficients for strings $\langle I_\alpha|E^\alpha_{pq}|J_\alpha\rangle$ $(\langle I_\beta|E_{pq}{}^\beta|J_\beta\rangle)$ can be stored in the computer memory in the form $J_\alpha[I_\alpha;\, pq]$ $(J_\beta[J_\beta\text{:p, q}])$. The perturbation calculation for three-body coupling coefficients, $\langle I_\alpha|E^\alpha_{pq,\,r,\,s}|J_\alpha\rangle$ $(\langle I_\beta|E_{pq}{}^\beta|J_\beta\rangle)$, for example, is performed as follows:

Loop over I_α
Make all non-zero $\langle I_\alpha|E^\alpha_{pq}|J_\alpha\rangle$ for I_α
Loop over I_β $[I_\alpha]$
Loop over t and u
If $\langle I_\beta|I_\alpha\, t,\, u] \neq 0$ and $|J_\beta\rangle(\langle I_\alpha;\, t,\, u] \in \{I_\beta[j_\alpha]\}$, *then*
do 3-body PT calculations for $\langle I_\alpha|E^\alpha_{pq,\,rs}|J_\alpha\rangle$ $(\langle I_\beta|E_{tu}{}^\beta|J_\beta\rangle)$
End loop t and u
End loop I_β $[I_\alpha]$
End loop I_α

The other terms can be computed similarly. The one- and two-body coupling coefficients computed in the same manner are used for the CI-based calculation for the internal terms. The vectors v_I in Eq. (34) are computed as s-vectors using strings.

A more efficient algorithm has been developed by using the α and β string spaces. Let us define the total space as a product of α and β string spaces. This method is called a string product space (SPS) SCF/PT method. Taking advantage of the independence of the α and β string spaces, the computational efforts can be dramatically reduced. For example, the perturbation calculation for three-body coupling coefficients. $\langle I_\alpha|E^\alpha_{pq,\,rs}|J_\alpha\rangle$ $(\langle I_\beta|E_{tu}^\beta|J_\beta\rangle)$ is simplified as follows:

Loop over I_α
Make all non-zero for $\langle I_\alpha|E^\alpha_{pq,\,rs}|J_\alpha\rangle$ for I_α
Loop over I_β $(=J_\beta)$
do 3-body PT calculations for $\langle I_\alpha|E^\alpha_{pq,\,rs}|J_\alpha\rangle$ $(\langle I_\beta|E_{tu}^\beta|J_\beta\rangle)$
End loop I_β
End loop I_α

Compared with the diagrammatic algorithm, the present scheme does not require loops over *t* and *u*. Usually, the α and β string spaces are spanned by the singly and doubly excited configurations, respectively. Thus, the total SPS consists of configurations up to quadruple excitations. The dimension of the reference function can be drastically reduced. The dimensions of CAS(14e, 14o) and CAS(18e, 18o) are 11 778 624 and 2 363 904 400, respectively, while the corresponding dimensions of SPS(14e, 14o) and SPS(18e, 18o) are 241 081 and 1 898 884. Thus, we can handle PT based on the (18e, 18o) reference functions. A numerical illustration shows that SPS-PT can describe the double-bond breaking process and treat several potential energy curves simultaneously. The excited states are also calculated very accurately by SPS-PT.

The convergence of the dynamical correlation is rather slow, and the accurate representation of the dynamical correlation requires high levels of excitations in the manyelectron wave function and high levels of polarization functions in a basis set. The situation is, however, quite different for nondynamical correlation. The nondynamical, near-degeneracy effect converges fairly smoothly with respect to both the one-electron basis function and the manyelectron wave function. This implies that the near degeneracy problem can be handled quite well even in a moderate function space. This supports the use of QCASSCF or MCSCF instead of CASSCF as a reference function in MRPT calculations. The QCAS and MCSCF methods are apparently quite poor when compared with the CASSCF, but the deficiency is largely overcome when the dynamical correlation is considered at the level of MRPT. MRPT can handle any state, regardless of charge, spin, or symmetry with surprisingly high and consistent accuracy, supporting our use of this method as our standard for treating small to medium-sized molecules. As is well known, ab initio computational effort depends heavily on the system, *N*. This dependence is of order N_4 for SCF and is of order N_5 and higher for MRPT.

The steep nonlinear cost of the conventional correlated methods has no physical origin; it is an artifact caused mainly by the use of canonical orbitals. Canonical orbitals, although conceptually and computationally convenient, destroy the local character of dynamical correlation. The development of alternative formulations based on local quantities is both feasible and desirable. The steep scaling can be reduced using local electron correlation methods. There has been continuing interest in local MRPT that uses the local character of the dynamical correlation. Research in this direction is now in progress.

9

Classification of Electronic States of Acetylene

The group theory represents a very powerful tool for the classification of electronic, vibrational and rotational states of molecules, particularly those possessing highly symmetrical nuclear configurations. A serious drawback of this approach is that the corresponding statements are generally of only qualitative nature. So, on the basis of symmetry considerations alone the number and type of electronic states arising by various populations of a set of molecular orbitals can be predicted, but not their energy ordering or the energy difference between them.

In the present chapter, it will be shown that also a lot of semiquantitative information can be obtained if pure group theory results are combined with some elementary quantum chemical considerations. This will be illustrated on the example of the acetylene molecule. This species was chosen for two reasons: firstly, it is interesting from the group theoretical point of view because of its relatively high symmetry (taking into account the small number of atoms) and the fact that the equilibrium geometries in its various electronic states belong to different point groups ($D_{\infty h}$, C_{2h}, C_{2v}, C^2, ...); secondly, the importance of this molecule may make the results of the present analysis interesting not only from a pure methodological point of view. The present approach is of course not new. Its roots represent the well known Mulliken-Walsh rules which enable a number of qualitative and even semiquantitative predictions concerning the geometry and energy ordering of electronic states of various classes of molecules to be made.

We shall shorten here this name to Walsh rules, in spite of the fact that particularly in Mulliken's hands this simple and elegant model has led to fascinating results; the reason for this is that there are also many other "Mulliken's rules" in this and related topics. The Walsh rules were derived in the precomputer era on the basis of simple MO-LCAO (molecular orbital as linear combination of atomic orbitals) considerations. The use of computers for solving the molecular Hartree-Fock (HF) equations has enabled the quantitative generation of the entities (molecular orbitals and the corresponding orbital energies) entering the Walsh model.

Although this quantification of the model introduces several new problems, not existing, or at least being hidden, from the pure qualitative point of view, there are at least two important advantages thereof: 1) it explains some important features not understandable on the basis of the qualitative MO theory alone; b) it represents an important link between naive concepts and the results of explicit *ab initio* calculations, the latter being of high (numerical) accuracy but are given in terms of very complex energy surfaces and corresponding wave functions, which obscures their interpretation.

MO Diagrams for HAAH Molecules

The behavior of MOs for the class of molecules with the formula HAAH, where A represents an atom belonging to the first row of the periodic table (e.g.,B, N, O), at the H-A and A-A stretching was discussed. In the present study are considered the angular (bending and torsional) dependence of the same quantities. To simplify the situation, it is assumed that all the bond lengths are kept fixed at their equilibrium values. At nuclear arrangements corresponding to these types of distortions with respect to the linear geometry ($D_{\infty h}$ point group) the molecule belongs to various point groups: C_{2h} (at the trans-bending), C_2 (cis-bending), C_2 (torsion at equal H-A-A bond angles). The correlation between the irreducible representations (irreps) of these point groups is given in Table I. For sake of completeness, the group C_s is also presented.

The choice of the axes is made according to the usual convention: For $D_{\infty h}$ the z-axis coincides with the molecular axis, for C_{2h}, C_2 and C_2 it represents the C_2 symmetry axis, and for C_s it is perpendicular to the symmetry (molecular) plane. In all cases except for $D_{\infty h}$ the y-axis is taken to lie along the A-A bond. For the discussion which follows it is convenient first to construct for each point group of interest the symmetrized linear combinations of the atomic orbitals (AO) building the minimal AO basis for the systems considered. The minimal AO basis consists of the following species:

$$1s_A\ ,\ 2s_A\ ,\ 2p_{xA} \equiv x_A,\ 2p_{yA} \equiv y_A,\ 2p_{zA} \equiv z_A,$$
$$1s_B,\ 2s_B,\ 2p_{xB} \equiv x_B\ ,\ 2py_B \equiv y_B,\ 2p_{zB} \equiv z_B,$$
$$1s_{HA} \equiv s_{HA},\ 1s_{HB} \equiv s_{HB}. \qquad (1)$$

In Eq. 1 different symbols (A,B) for the two heavy atoms and the two hydrogens (HA, HB) are introduced. In this new notation the HAAH molecule reads HAABHB. The symmetric and antisymmetric linear combination of pairs of these twelve AOs build the bases for irreducible representations of the point groups in question, as given in Table II. Note that the symmetry species appearing in the same row of Table II do not necessarily correlate with one another (e.g., π_u, b_u+b_u, b_1+b_2, b+b, corresponding to the two-dimensional space spanned by x_A+x_B and y_A+y_B), because of the different meaning of the x, y, and z axes for the different point groups. In the C_s point group each individual AO, except for z_A and z_B belongs to a', the latter two being of a'' symmetry.

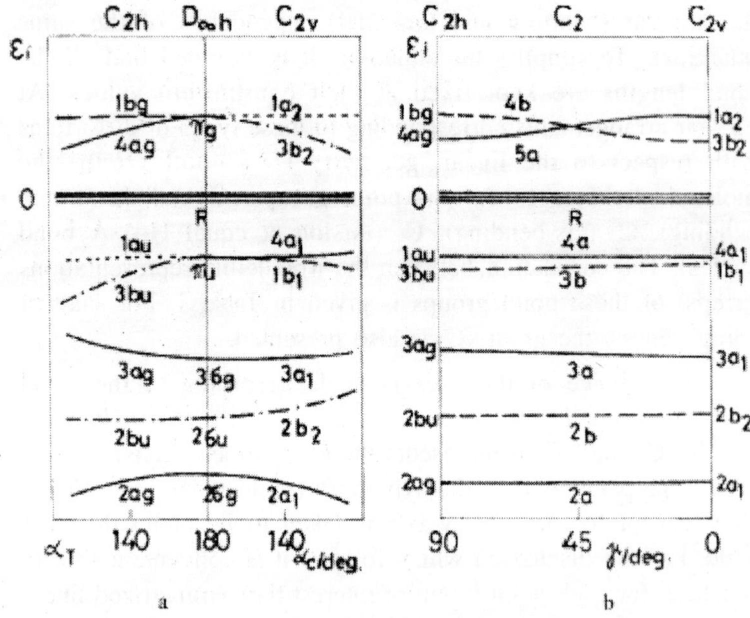

Fig. 1. The dependence of MO energies on variation of geometry in symmetric tetraatomic molecules HAAH. 1a: Trans- and cis-bending. Solid lines denote species of ag $_{(C2h}$ point group) and a_1 (C_2v group) symmetry, dotted lines a_u and a_2, dashed lines b_g and b_1, dash-dotted lines b_u and b_2. 1b: Torsional dependence of MO energies.

Besides these twelve ("valence") AOs, the s, p, ... Rydberg orbitals will also be considered, because the latter are involved in the series of experimentally observed excited electronic states of the acetylene molecule. These orbitals are characterized by large spatial extension and closely resemble the orbitals of an isolated atom. They can all be assumed to be centered at the mid-point of the molecule and each of them transforms separately according to the total-symmetrical irrep (s-species), like the *x*, *y*, *z* coordinates (p_x, p_y, p_z Rydberg orbitals, respectively), etc., in each of the point groups considered.

Table I

Correlation of the species of a linear molecule ($D_{\infty h}$ point group) with those of a molecule of lower symmetry

$D_{\infty h}$	C_{2h}	C_2	C_2	C_s
Σ_g^+	Ag	A_1	A	A'
Σ_g^-	Bg	B_1	B	A"
Π_g	$A_g + B_g$	$A_2 + B_2$	A+ B	A'+ A"
Δ_g	$A_g + B_g$	$A_1 + B_1$	A+ B	A'+ A"
Φ_g	$A_g + B_g$	$A_2 + B_2$	A+ B	A'+ A"
...	...	...	...	...
Σ_u^+	B_u	B_2	B	A'
Σ_u^-	A_u	A_2	A	A"
Π_u	$A_u + B_u$	$A_1 + B_1$	A+ B	A'+ A"
Δ_u	$A_u + B_u$	$A_2 + B_2$	A+ B	A'+ A"
Φ_u	$A_u + B_u$	$A_1 + B_1$	A+ B	A'+ A"
...	...	...	...	...

Table II

Classification of atomic orbitals according to the irreducible representations of the point groups of interest

	$D_{\infty h}$	C_{2h}	C_2	C_s
$1s_A + 1s_B$	σ_g	a_g	a_1	a
$1s_A - 1s_B$	σ_u	b_u	b_2	b
$2s_A + 2s_B$	σ_g	a_g	a_1	a
$2s_A - 2s_B$	σ_u	b_u	b_2	b
$x_A + x_B$	π_u	b_u	b_1	b
$y_A + y_B$	σ_h	b_u	b_2	b
$z_A + z_B$	σ_u	a_u	a_1	a
$x_A - x_B$	π_g	a_g	a_2	a

(Contd...)

$y_A - y_B$		a_1	a	
$z_A - z_B$	σ_g	b_g	b_2	b
$H_A + H_B$	σ_g	a_g	a_1	a
$H_A - H_B$	σ_u	b_u	b_2	b

The dependence of the energies of the low-energy orbitals up to $1\pi_g$ (except for the $1\sigma_g$ and $1\sigma_u$ which are characterized by much lower energies than those of all other species) on the bending coordinates is presented in Fig. 1a. It is extracted from HF calculations on several HAAH molecules and is in most, but not all instances, qualitatively the same as that predicted in the original studies by Mulliken and Walsh. The data of Fig. 1a is of semiquantitative nature in the sense that the energy ordering, increase or decrease of the curves are reproduced quite correctly, while the energy differences between the curves are arbitrary.

The composition of the MOs for the linear nuclear arrangement will be discussed first. The two MOs with the lowest energy, $1\sigma_g$ and $1\sigma_u$ (not shown in Fig. 1a), are built by the symmetric and antisymmetric linear combinations of the 1s AOs of the A, B atoms. The electrons populating these orbitals are mainly localized at the heavy nuclei and do not contribute significantly to the binding of the atoms in the molecule. The next two orbitals in order of increasing energy are $2\sigma_g$ and $2\sigma_u$, composed predominately by the symmetric and antisymmetric linear combinations of the 2s A, B orbitals, respectively, with a small admixture of the hydrogen 1s AOs. They are followed by $3\sigma_g$, involving the antisymmetric linear combination of the 2pz A, B orbitals directed along the molecular axis (2pσ) and the symmetric linear combination of the hydrogen 1s AOs. The next orbital is $1\pi_u$, composed of the symmetric linear combination of the p orbitals of the A,B atoms perpendicular to the molecular axis ($2p_x$, $2p_y$ = $2p\pi$). If the AO basis employed did not involve the Rydberg-type species, the lowest-lying orbitals not occupied in the ground state of C_2H_2 ("virtual orbital") would be $1\pi_g$, built by the symmetric linear combinations of the 2pπ A, B

AOs. It would be followed by the $3\sigma_u$ orbital, etc. However, the Rydberg orbitals present in the AO basis employed enter into a branch of non-bonding MOs, characterized with energies close to zero, and lying thus between $1\pi_u$ and $1\pi_g$. Let us note that discrete energies of virtual orbitals are actually artificial; if the AO basis were infinite, the spectrum of virtual orbitals would be continuous. The behaviour of the orbital energies upon bending is governed by several effects. First of all, because of the reduced symmetry, all the species doubly degenerate in the linear geometry (π, δ, ...) split into two components.

The second important effect is that in the lower-symmetry groups more AOs are generally involved in a MO belonging to a particular irrep, which can contribute to a lowering of its energy - an example for this are the $2a_g$ and $2a_1$ orbitals compared with $2\sigma_g$ to which they correlate in the linear nuclear arrangement. The overlap between the AOs belonging to a hydrogen and a heavy atom within a MO can become more (as, for example, in $3b_u$) or less (e.g., in $3a_g$ and $3a_1$) pronounced upon bending; in the first case such a MO is stabilized, in the second case it is destabilized upon bending.

The next effect influencing the geometry dependence of orbital energies is mutual repelling of MOs of the same symmetry and similar energy (as e.g., $3a_1$ and $4a_1$). Finally, the composition of MOs and consequently their energy is influenced by the requirement for their mutual orthogonality. The form of the curves presented in Fig. 1a can be interpreted more or less straightforwardly by taking into account all these facts. The behavior of the MOs involving also Rydberg-type AOs is clearly dominated by the properties of these species; these MOs show no significant change in composition and energy with variation of the geometry.

In. Fig. 1b are displayed the orbital energies as functions of the torsion angle at a particular value for the bending angle $\angle H_A\text{-}A\text{-}B = \angle A\text{-}B\text{-}H_B$ (of, say 120°). The torsion angle γ is defined as half the angle between the H_AAB and ABH_B planes.

This means that $\gamma = 0$ for cis-planar geometry, and $\gamma = \pi/2$ for trans-planar nuclear arrangement. Only the 5a and 4b orbitals show significant change upon torsion. It is easy to explain: 5a orbital correlates at trans-planar geometry ($\gamma = \pi/2$) with the $4a_g$ species; this MO is predominantly built by the antisymmetric linear combination of the p_x orbitals of the heavy atoms (lying in the molecular plane), but is significantly admixed by the symmetric linear combination of the hydrogen 1s function. The overlap between these functions decreases with decreasing torsional angle. At *cis*-planar geometry, the 5a orbital correlates with $1a_2$, the latter does not involve the hydrogen AOs for symmetry reasons. A consequence of these facts is that the energy of the 5a orbital decreases on changing of the molecular geometry from *trans*-planar towards *cis*-planar. The behavior of the 4b species is just the opposite.

Vertical Electronic Spectrum of Acetylene

In the ground electronic state, the acetylene molecule is linear. The electronic configuration of the ground state corresponds to the distribution of fourteen electrons among the lowest-energy MOs available: $1\sigma_g^2\, 1\sigma_u^2\, 1\sigma_g^2\, 1\sigma_u^2\, 1\sigma_g^2\, 1\pi_u^4$. The symmetry of this state is thus $1\Sigma_g^+$.

The lowest-lying excited state of acetylene corresponds to a one-electron excitation from the highest MO populated in the ground state $1\pi_u \equiv \pi_u$, into the lowest lying unpopulated ("virtual") MO, $1\pi_g \equiv \pi_g$ ($\pi_u{}^3\pi_g$ configurations). When considering the linear nuclear arrangement, it is convenient to use their linear combinations which are eigenfunctions of the projection of the electronic angular momentum operator onto the molecular axis, L_z. All the electronic configurations which are considered in the present study have two electrons in open shells, i.e., in (spatial) orbitals populated with a single electron (at linear geometry there will actually be situations where three electrons occupy a π orbital - such a case can and shall be treated as a single "hole" in this MO). Since the electrons occupying closed shells do not contribute to the molecular angular (spatial and spin) momentum, the z-components of the angular momentum and spin operator of the molecule can be written in the form

$$L_z = l_{z1} + l_{z2} = -i\left(\frac{\partial}{\partial\phi_1} + \frac{\partial}{\partial\phi_2}\right); S_z = s_{z1} + s_{z2} \qquad (2)$$

(atomic units, $m_e \equiv 1$, $e \equiv 1$, $h \equiv 1$ are used throughout this chapter) where indices 1 and 2 denote the electrons which can be outside the closed shells, and ϕ_1, ϕ_2 represent their azimuthal angles. To obtain the required components of the orbitals one starts with the p AOs of the heavy atoms (A, B) expressed in polar coordinates,

$$2p_{xA} \equiv x_A = f(\rho, z_A) \cos \phi,\ 2p_{xB} \equiv x_B = f(\rho, z_B) \cos \phi$$
$$2p_{yA} \equiv y_A = f(\rho, z_A) \sin \phi,\ 2p_{yB} \equiv y_B = \phi(\rho, z_B) \sin \phi \quad (3)$$

where the symbol z_A (z_B) indicates that the corresponding function is centered at the nucleus A(B). It follows that

$$x_A + x_B = f_u \cos \phi,\ x_A - x_B = f_g \cos \phi,$$
$$y_A + y_B = f_u \sin \phi,\ y_A - y_B = f_g \sin \phi, \qquad (4)$$

where

$$f_u \equiv f(\pi, z_A) + f(\rho, z_B),\ f_g \equiv f(\rho, z_A) - f(\rho, z_B). \qquad (5)$$

Now the linear combinations of (4) are built,

$$\pi_u = \frac{1}{2\sqrt{1+S}}[(x_A + x_B) + i(y_A + y_B)] = f_u e^{i\phi}$$

$$\bar{\pi}_u = \frac{1}{2\sqrt{1+S}}[(x_A + x_B) - i\,(y_A + y_B)] = f_u\, e^{-i\phi}$$
$$\pi_g = \frac{1}{2\sqrt{1-S}}[(x_A - x_B) + i\,(y_A - y_B)] = f_g\, e^{i\phi} \qquad (6)$$
$$\bar{\pi}_g = \frac{1}{2\sqrt{1-S}}[(x_A - x_B) - i\,(y_A - y_B)] = f_g\, e^{-i\phi}$$

It is assumed that AOs are real and normalized (but, of course, not mutually orthogonal in the general case); S is the overlap integral,

$$S \equiv \int x_A\, x_B\, dr \equiv \int y_A\, y_B\, dr. \qquad (7)$$

There are 16 Slater detrminants corresponding to the $\pi u^3\, \pi_g$ configurations, which are in accordance with the Pauli principle:

$$D_1 = |\bar{\pi}_u\alpha, \pi_g\alpha|,\quad D_2 = |\bar{\pi}_u\alpha, \pi_g\beta|,$$
$$D_3 = |\bar{\pi}_u\alpha, \bar{\pi}_g\alpha|,\quad D_4 = |\bar{\pi}_u\alpha, \bar{\pi}_g\beta|,$$
$$D_5 = |\bar{\pi}_u\beta, \pi_g\alpha|,\quad D_6 = |\bar{\pi}_u\beta, \pi_g\beta|$$
$$D_7 = |\bar{\pi}_u\beta, \bar{\pi}_g\alpha|,\quad D_8 = |\bar{\pi}_u\beta, \bar{\pi}_g\beta|,$$
$$D_9 = |\pi_u\alpha, \pi_g\alpha|,\quad D_{10} = |\pi_u\alpha, \pi_g\beta|,$$
$$D_{11} = |\pi_u\alpha, \bar{\pi}_g\alpha|,\quad D_{12} = |\pi_u\alpha, \bar{\pi}_g\beta|,$$
$$D_{13} = |\pi_u\beta, \pi_g\alpha|,\quad D_{14} = |\pi_u\beta, \pi_g\beta|,$$
$$D_{15} = |\pi_u\beta, \bar{\pi}_g\alpha|,\quad D_{16} = |\pi_u\beta, \bar{\pi}_g\beta|, \qquad (8)$$

The shortened notation for the Slater determinants will be employed, e.g.,

$$|\pi\alpha, \bar{\pi}\beta| \equiv \frac{1}{\sqrt{2}} \begin{vmatrix} \pi(1)\alpha(1) & \bar{\pi}(1)\beta(1) \\ \pi(2)\alpha(2) & \bar{\pi}(2)\beta(2) \end{vmatrix} \quad (9)$$

The electronic Hamiltonian of a linear molecule commutes with the operators L_z, S^2 and Sz, and thus the quantum numbers corresponding to these operators, A, S and M_S are "good" quantum numbers. All the Slater determinants (8) are eigenfunctions of L_z and S_z but generally not of S^2. Additionally, the Hamiltonian commutes with the operator for permutations of identical nuclei (A, B and H_A, H_B), as well as with the operator σ_v corresponding to the reflection of all electronic spatial coordinates in the planes crossing one another along the molecular axis. Thus, a correct electronic wave function is labeled by the quantum number *g* or *u,* according to its behavior upon reflection in the nuclear inversion centrum, and the wave function for a Σ (Λ = 0) electronic states also by + if it is invariant upon reflection in σ_v and by - if it changes sign under this operation. From the Slater determinants (8) the following spectroscopic states are constructed:

$$\begin{gathered}
{}^1\Delta_u = \frac{1}{\sqrt{2}}(D_{10} - D_{13}) = {}^1\Phi_2\,\Theta^1, \\
{}^1\bar{\Delta}_u = \frac{1}{\sqrt{2}}(D_4 - D_7) = {}^1\Phi_{-2}\,\Theta^1; \\
{}^3\Delta_u\,(M_S = 1) = D_9 = {}^3\Phi_2\,\Theta^3{}_1, \\
{}^3\Delta_u\,(M_S = 0) = \frac{1}{\sqrt{2}} = (D_{10} + D_{13}) = {}^3\Phi_2\,\Theta^3{}_0, \\
{}^3\Delta_u\,(M_S = -1) = D_{14} = {}^3\Phi_2\,\Theta^3{}_{-1}; \\
{}^3\bar{\Delta}_u\,(M_S = 1) = D_3 = {}^3\Phi_{-2}\,\Theta^3{}_1, \\
{}^3\bar{\Delta}_u\,(M_S = 0) = \frac{1}{\sqrt{2}}(D_4 + D_7) = {}^3\Phi_{-2}\,\Theta^3{}_0,
\end{gathered} \quad (10)$$

$$^3\overline{\Delta}_u\,(M_S=-1)=D_8={}^3\Phi_{-2}\,\Theta^3{}_{-1},$$

$$^1\Sigma_u{}^+=\frac{1}{2}\,(D_2-D_5+D_{12}-D_{15})={}^1\Phi^+\,\Theta^1,$$

$$^3\Sigma_u{}^+\,(M_S=1)=\frac{1}{\sqrt{2}}\,(D_1+D_{11})={}^3\Phi^+\,\Theta^3{}_1,$$

$$^3\Sigma_u{}^+\,(M_S=0)=\frac{1}{2}\,(D_2+D_5+D_{12}+D_{15})={}^3\Phi^+\,\Theta^3{}_0,$$

$$^3\Sigma_u{}^+\,(M_S=-1)=\frac{1}{\sqrt{2}}\,(D_6+D_{16})={}^3\Phi^+\,\Theta^3{}_{-1};$$

$$^3\Sigma_u{}^-=\frac{1}{2}\,(D_2-D_5-D_{12}+D_{15})={}^1\Phi^-\,\Theta^1,$$

$$^3\Sigma_u{}^-\,(M_S=1)=\frac{1}{\sqrt{2}}\,(D_1-D_{11})={}^3\Phi^-\,\Theta^3{}_1,$$

$$^3\Sigma_u{}^-\,(M_S=0)=\frac{1}{2}\,(D_2+D_5-D_{12}-D_{15})={}^3\Phi^-\,\Theta^3{}_0,$$

$$^3\Sigma_u{}^-\,(M_S=-1)=\frac{1}{\sqrt{2}}\,(D_6-D_{16})={}^3\Phi^-\,\Theta^3{}_{-1}.$$

On the left-hand side Δ stands for the species with $\Lambda = 2$, $\overline{\Delta}$ for those corresponding to $\Lambda = -2$. The components of a triplet state are denoted by the value of M_s given in parentheses. The functions appearing in Eqs. (10) are defined as

$$^1\Phi_2\equiv\frac{1}{\sqrt{2}}\,[\pi_u(1)\,\pi_g(2)+\pi_g(1)\,\pi_u(2)]\equiv\frac{1}{\sqrt{2}}[\pi_u\pi_g+\pi_g\pi_u]=$$

$$=\frac{1}{\sqrt{2}}\,[f_u(1)\,f_g(2)+f_g(1)\,f_u(2)]\,\exp[i(\phi_1+\phi_2)]\equiv$$

$$\frac{1}{\sqrt{2}}\,[f_uf_g+f_gf_u]\,\exp[i(\phi_1+\phi_2)],$$

$$^1\Phi_{-2}=\frac{1}{\sqrt{2}}\,[\overline{\pi}_u\overline{\pi}_g+\overline{\pi}_g\overline{\pi}_u]=\frac{1}{\sqrt{2}}\,[f_uf_g+f_gf_u]\,\exp[-i(\phi_1+\phi_2)\,],$$

$$^3\Phi_2=\frac{1}{\sqrt{2}}\,[\pi_u\pi_g-\pi_g\pi_u]=\frac{1}{\sqrt{2}}\,[f_uf_g-f_gf_u]\,\exp[i(\phi_1+\phi_2)\,],$$

(11)

$$^3\Phi_{-2}=\frac{1}{\sqrt{2}}\,[\bar{\pi}_u\bar{\pi}_g-\bar{\pi}_g\bar{\pi}_u]=\frac{1}{\sqrt{2}}[f_uf_g-f_gf_u]\,\exp[-i(\phi_1+\phi_2)\,],$$

$$^1\Phi^+=\frac{1}{2}\,[\pi_u\bar{\pi}_g+\bar{\pi}_g\pi_u+\bar{\pi}_u\pi_g+\pi_g\bar{\pi}_u]=[f_uf_g+f_gf_u]\,\cos(\phi_1-\phi_2)\,,$$

$$^3\Phi^+=\frac{1}{2}\,[\pi_u\bar{\pi}_g-\bar{\pi}_g\pi_u+\bar{\pi}_u\pi_g-\pi_g\bar{\pi}_u]=[f_uf_g-f_gf_u]\,\cos(\phi_1-\phi_2)\,,$$

$$^1\Phi^-=\frac{1}{2}\,[\pi_u\bar{\pi}_g+\bar{\pi}_g\pi_u-\bar{\pi}_u\pi_g-\pi_g\bar{\pi}_u]=\,i\,[f_uf_g-f_gf_u]\,\sin(\phi_1-\phi_2)\,,$$

$$^3\Phi^-=\frac{1}{2}\,[\pi_u\bar{\pi}_g-\bar{\pi}_g\pi_u-\bar{\pi}_u\pi_g+\pi_g\bar{\pi}_u]=i\,[f_uf_g+f_gf_u]\,\sin(\phi_1-\phi_2)\,,$$

and

$$\begin{aligned}\Theta^1 &\equiv\frac{1}{\sqrt{2}}\,[\alpha(1)\,\beta(2)-\beta(1)\alpha(2)\,]\equiv\frac{1}{\sqrt{2}}\,(\alpha\beta-\beta\alpha)\\ \Theta^3{}_1 &=\alpha(1)\,\alpha(2)\equiv\alpha\alpha,\\ \Theta^3{}_0 &=\frac{1}{\sqrt{2}}\,[\alpha(1)\,\beta(2)+\beta(1)\,\alpha(2)]\equiv\frac{1}{\sqrt{2}}\,(\alpha\beta+\beta\alpha)\\ \Theta^3{}_{-1} &=\beta(1)\,\beta(2)\equiv\beta\beta\,.\end{aligned}\qquad(12)$$

Now the approximate expectation values for the electronic Hamiltonian, corresponding to the wave functions defined by Eqs. (10), will be derived. It is assumed that the contribution from the closed shells is the same in all cases, and that the relative ordering of the states considered can be obtained by computing the mean value of the Hamiltonian for two electrons,

$$H=h_1+h_2+\frac{1}{r_{12}}\qquad(13)$$

where h_1 and h_2 are one-electron operators and r_{12} is the distance between the electrons 1 and 2. The contribution from the one-electron operators h_1 and h_2 is for all the present cases ($\pi_u\pi_g$ configurations)

$$<\pi_u|h/_{12}|\pi_u>+<\pi_g|h_{1/2}|\pi_g> \equiv \varepsilon_u + \varepsilon_g \quad (14)$$

The mean values for the two-electron operator $1/r_{12}$ are

$$^1\Delta_u : J_{ug} + K_{ug} =$$

$$\frac{1}{2(1-S^2)}\{(x_A x_A \mid x_A x_A) - (x_A x_B \mid x_A x_B) +$$

$$(x_A x_A \mid y_A y_A) - (x_A x_B \mid y_A y_B) +$$

$$+ (x_A y_B \mid x_A y_B) - (x_A y_B \mid x_B y_A)\},$$

$$^3\Delta_u : J_{ug} - K_{ug} =$$

$$\frac{1}{2(1-S^2)}\{(x_A x_A \mid x_B x_B) - (x_A x_B \mid x_A x_B) +$$

$$(x_A x_A \mid y_B y_B) - (x_A x_B \mid y_A y_B) -$$

$$-(x_A y_B \mid x_A y_B) + (x_A y_B \mid x_B y_A)\}, \quad (15)$$

$$^1\Sigma_u^+ : J_{u\bar{g}} + K_{u\bar{g}} + \frac{1}{1-S^2}[(\pi_u \bar{\pi}_u \mid \bar{\pi}_g \pi_g) +$$

$$(\pi_u \pi_g \mid \bar{\pi}_g \bar{\pi}_u)] =$$

$$= \frac{1}{1-S^2}\{(x_A x_A \mid x_A x_A) - (x_A x_B \mid x_A x_B) +$$

$$(x_A y_A \mid x_A y_A) - (x_A y_B \mid x_A y_B)\},$$

$${}^3\Sigma_u^+ : J_{u\bar{g}} - K_{u\bar{g}} + \frac{1}{1-S^2}[(\pi_u \bar{\pi}_u \mid \bar{\pi}_g \pi_g) - (\pi_u \pi_g \mid \bar{\pi}_g \bar{\pi}_u)] =$$

$$= \frac{1}{1-S^2}\{(x_A x_A \mid x_B x_B) - (x_A x_B \mid x_A x_B) +$$

$$(x_A y_A \mid x_B y_B) - (x_A y_B \mid x_B y_A)\},$$

$${}^1\Sigma_u^- : J_{u\bar{g}} + K_{u\bar{g}} - \frac{1}{1-S^2}[(\pi_u \bar{\pi}_u \mid \bar{\pi}_g \pi_g) + (\pi_u \pi_g \mid \bar{\pi}_g \bar{\pi}_u)] =$$

$$= \frac{1}{1-S^2}\{(x_A x_A \mid y_B y_B) - (x_A x_B \mid y_A y_B) -$$

$$(x_A y_A \mid x_B y_B) + (x_A y_B \mid x_A y_B)\},$$

$$= \frac{1}{1-S^2}\{(x_A x_A \mid y_A y_A) - (x_A x_B \mid y_A y_B) -$$

$$(x_A y_A \mid x_A y_A) + (x_A y_B \mid x_B y_A)\}.$$

$${}^3\Sigma_u^- : J_{u\bar{g}} - K_{u\bar{g}} - \frac{1}{1-S^2}[(\pi_u \bar{\pi}_u \mid \bar{\pi}_g \pi_g) - (\pi_u \pi_g \mid \bar{\pi}_g \bar{\pi}_u)]$$

In Eq. (15), the "chemists' notation" for four-center integrals is employed,

$$\iint a^*(1) b^*(2) \frac{1}{r_{12}} c(1) d(2) d\tau_1 d\tau_2 \equiv (a\,c \mid b\,d), \qquad (16)$$

the Coulomb and exchange integrals are introduced

$$J_{ug} \equiv (u\,u \mid g\,g), \quad K_{ug} \equiv (u\,g \mid g\,u),$$

$$\cdot \mid a_u\,\alpha, b_g\,\alpha \mid, D_4{}^{T} = \mid a_u\,\alpha, b_g\,\beta \mid, \tag{17}$$

$$D_5{}^{T} = \mid a_u\,\beta, a_g\,\alpha \mid, D_6{}^{T} = \mid a_u\,\beta, a_g\,\beta \mid, D_7{}^{T} =$$

(where u denotes π_u and g π_g), and the matrix elements expressed also in terms of the atomic basis functions.

The values for the four-center integrals can be estimated by means of the Mulliken formula

$$(a\,b \mid c\,d) \approx \frac{1}{4} S_{ab} S_{cd} [(a\,a \mid c\,c) + (b\,b \mid c\,c) + (a\,a \mid d\,d) + (b\,b \mid d\,d)], \tag{18}$$

where S_{ab} and S_{cd} are overlap integrals for the orbitals a, b, and c, d, respectively. An analysis of the expressions (15) leads to the following ordering of the acetylene excited electronic states in order of increasing energy:

$$^3\Sigma_u{}^+, {}^3\Delta_u, {}^3\Sigma_u{}^-, {}^1\Sigma_u{}^-, {}^1\Delta_u, {}^1\Sigma_u{}^+. \tag{19}$$

All these states except for $1\Sigma_u{}^+$ (lying at very high energy) represent the lowest-lying excited species of the acetylene molecule.

The excited electronic states corresponding to the $\pi_u{}^3\pi_g$ electronic configurations are followed by a series of Rydberg-type states arising by excitations out of the π_u orbital into the orbitals involving the Rydberg AOs. The energy positions of these species generally match very reasonably the formula

$$T_R = IP - R/(n - \delta)^2 \tag{20}$$

where T_R is the term value of the state in question, *IP* the ionization potential of the molecule (11.4 eV), *R* the Rydberg constant, n the principal quantum number of the Rydberg state, and δ the quantum defect (usually assumed to be $\delta = 1.0$ for s series, $\delta = 0.4/0.5$ for p, and $\delta = 0.0/0.1$ for d states). The only exception represent the lowest-lying states of both singlet and triplet multiplicity (corresponding to the $1\pi_u \rightarrow 3s_R$ electron excitation) the vertical energies of which differ significantly from those obtained by means of formula 20, indicating that these species are of mixed Rydberg-valence character. The Rydberg states of acetylene converge towards the ground state, $X^2\Pi_u$, of the $C_2H_2^+$ ion.

Trans-bending Potential Curves

At *trans*-bent nuclear arrangements (point group C_{2h}), the π_u and π_g orbitals split into $3b_u + 1a_u$ and $4a_g + 1b_g$, respectively. One-electron excitations out of $3b_u$ ($\equiv b_u$) or $1a_u$ ($\equiv a_u$) into $4a_g$ ($\equiv a_g$) or $1b_g$ ($\equiv b_g$) lead to the electronic species represented by the following sixteen Slater determinantsm

$$D_1^T = |a_u\,\alpha, a_g\,\alpha|, D_2^T = |a_u\,\alpha, a_g\,\beta|, D_3^T =$$

$$|a_u\,\beta, b_g\,\alpha|, D_8^T = |a_u\,\beta, b_g\,\beta|,$$

$$D_9^T = |b_u\,\alpha, a_g\,\alpha|, D_{10}^T = |b_u\,\alpha, a_g\,\beta|, D_{11}^T = \tag{21}$$

$$|b_u\,\alpha, b_g\,\alpha|, D_{12}^T = |b_u\,\alpha, b_g\,\beta|,$$

$$D_{13}^T = |b_u\,\beta, a_g\,\alpha|, D_{14}^T = |b_u\,\beta, a_g\,\beta|, D_{15}^T =$$

$$|b_u\,\beta, b_g\,\alpha|, D_{16}^T = |b_u\,\beta, b_g\,\beta|.$$

These Slater determinants are combined into "spectroscopically correct" electronic states, i.e., species being eigenfunctions of the spin operators S^2 and S_z and belonging to a particular irrep of the C_{2h} point group as

$$
\begin{aligned}
1\,{}^1\mathrm{A_u} &= \frac{1}{\sqrt{2}}(D_2{}^\mathrm{T} - D_5{}^\mathrm{T}) = \frac{1}{\sqrt{2}}(\mathrm{a_u a_g} + \mathrm{a_g a_u})\,\Theta^1, \\
2\,{}^1\mathrm{A_u} &= \frac{1}{\sqrt{2}}(D_{12}{}^\mathrm{T} - D_{15}{}^\mathrm{T}) = \frac{1}{\sqrt{2}}(\mathrm{b_u b_g} + \mathrm{b_g b_u})\,\Theta^1; \\
1\,{}^3\mathrm{A_u}\,(M_s = 1) &= D_1{}^\mathrm{T} = \frac{1}{\sqrt{2}}(\mathrm{a_u a_g} - \mathrm{a_g a_u})\,\Theta_1{}^3, \\
1\,{}^3\mathrm{A_u}\,(M_s = 0) &= \frac{1}{\sqrt{2}}(D_2{}^\mathrm{T} + D_5{}^\mathrm{T}) = \frac{1}{\sqrt{2}}(\mathrm{a_u a_g} - \mathrm{a_g a_u})\,\Theta_0{}^3, \\
1\,{}^3\mathrm{A_u}\,(M_s = -1) &= D_6{}^\mathrm{T} = \frac{1}{\sqrt{2}}(\mathrm{a_u a_g} - \mathrm{a_g a_u})\,\Theta_{-1}{}^3; \\
2\,{}^3\mathrm{A_u}\,(M_s = 1) &= D_{11}{}^\mathrm{T} = \frac{1}{\sqrt{2}}(\mathrm{b_u b_g} - \mathrm{b_g b_u})\,\Theta_1{}^3, \\
2\,{}^3\mathrm{A_u}\,(M_s = 0) &= \frac{1}{\sqrt{2}}(D_{12}{}^\mathrm{T} + D_{15}{}^\mathrm{T}) = \frac{1}{\sqrt{2}}(\mathrm{b_u b_g} - \mathrm{b_g b_u})\,\Theta_0{}^3, \\
2\,{}^3\mathrm{A_u}\,(M_s = -1) &= D_{16}{}^\mathrm{T} = \frac{1}{\sqrt{2}}(\mathrm{b_u b_g} - \mathrm{b_g b_u})\,\Theta_{-1}{}^3, \\
1\,{}^1\mathrm{B_u} &= \frac{1}{\sqrt{2}}(D_4{}^\mathrm{T} - D_7{}^\mathrm{T}) = \frac{1}{\sqrt{2}}(\mathrm{a_u b_g} + \mathrm{b_g a_u})\,\Theta^1, \qquad (22) \\
2\,{}^1\mathrm{B_u} &= \frac{1}{\sqrt{2}}(D_{10}{}^\mathrm{T} - D_{13}{}^\mathrm{T}) = \frac{1}{\sqrt{2}}(\mathrm{b_u a_g} + \mathrm{a_g b_u})\,\Theta^1, \\
1\,{}^3\mathrm{B_u}\,(M_s = 1) &= D_3{}^\mathrm{T} = \frac{1}{\sqrt{2}}(\mathrm{a_u b_g} - \mathrm{b_g a_u})\,\Theta_1{}^3, \\
1\,{}^3\mathrm{B_u}\,(M_s = 0) &= \frac{1}{\sqrt{2}}(D_4{}^\mathrm{T} + D_7{}^\mathrm{T}) = \frac{1}{\sqrt{2}}(\mathrm{a_u b_g} - \mathrm{b_g a_u})\,\Theta_0{}^3; \\
1\,{}^3\mathrm{B_u}\,(M_s = -1) &= D_8{}^\mathrm{T} = (\mathrm{a_u b_g} - \mathrm{b_g a_u})\,\Theta_{-1}{}^3, \\
2\,{}^3\mathrm{B_u}\,(M_s = 1) &= D_9{}^\mathrm{T} = \frac{1}{\sqrt{2}}(\mathrm{b_u a_g} - \mathrm{a_g b_u})\,\Theta_1{}^3,
\end{aligned}
$$

$$2\,^3B_u\,(M_s=0)=\frac{1}{\sqrt{2}}\,(D_{10}{}^T+D_{13}{}^T)=\frac{1}{\sqrt{2}}\,(b_ua_g-a_gb_u)\,\Theta_0{}^3,$$

$$2\,^3B_u\,(M_s=-1)=D_{14}{}^T=\frac{1}{\sqrt{2}}(b_ua_g-a_gb_u)\,\Theta_{-1}{}^3.$$

The MOs in the $D_{\infty h}$ point group are connected with their C2h counterparts by the relations

$$\pi_u=\frac{1}{\sqrt{2}}\,(b_u+i\,a_u),\ \ \bar{\pi}_u=\frac{1}{\sqrt{2}}\,(b_u-i\,a_u),$$

$$\pi_u=\frac{1}{\sqrt{2}}\,(a_g+i\,b_g),\ \ \bar{\pi}_u=\frac{1}{\sqrt{2}}\,(a_g-i\,b_g). \qquad (23)$$

By using (23), the correlation between the electronic states of the two point groups considered can be derived. It reads

$$\frac{1}{\sqrt{2}}\,(^1\Delta_u+{}^1\bar{\Delta}_u)\rightarrow\frac{1}{\sqrt{2}}(2^1B_u-1^1B_u),$$

$$\frac{1}{\sqrt{2}}\,(^1\Delta_u-{}^1\bar{\Delta}_u)\rightarrow\frac{i}{\sqrt{2}}(2^1A_u-1^1A_u),$$

$$\frac{1}{\sqrt{2}}\,(^3\Delta_u+{}^3\bar{\Delta}_u)\rightarrow\frac{1}{\sqrt{2}}(1^3B_u-2^3B_u),$$

$$\frac{1}{\sqrt{2}}\,(^3\Delta_u-{}^3\bar{\Delta}_u)\rightarrow\frac{i}{\sqrt{2}}(2^3A_u+1^3A_u),$$

$$^1\Sigma_u{}^+\rightarrow\frac{1}{\sqrt{2}}\,(2^1B_u+1^1B_u), \qquad (24)$$

$$^1\Sigma_u{}^-\rightarrow\frac{i}{\sqrt{2}}\,(1^1A_u-2^1A_u),$$

$$^3\Sigma_u^+ \rightarrow \frac{1}{\sqrt{2}}(1^3B_u - 2^3B_u),$$

$$^3\Sigma_u^- \rightarrow \frac{i}{\sqrt{2}}(1^3A_u - 2^3A_u).$$

The geometry of the electronic states of acetylene in terms of the energy change of the MOs involved in the corresponding wave functions upon trans-bending are now discussed. Only the singlet electronic states will be considered - the analysis of triplet species can be carried out in a completely analogous way. At the linear nuclear arrangement, the electronic configuration of the ground electronic state of acetylene, $X^1\Sigma_g^+$ is ... π_u^4. At *trans*-bent geometries it becomes ... $3b_u^2 1a^{u2}$. The effect of decreasing the energy of the $3bu^2$ orbital upon bending is outweighed by the strong energy increase of the 3agMO (also doubly occupied), resulting in a linear equilibrium geometry of the $X^1\Sigma_g^+$ state. The correlation scheme given by Eqs. (24) shows that the first excited singlet state, $^1\Sigma_u^-$, correlates with the antisymmetric linear combination of the 1^1A_u and 2^1A_u species of the C_{2h} point group, with coefficients of exactly equal magnitude ($1/\sqrt{2}$). However, already at small distortions from linearity, the 1^1A_u component becomes dominant. The $^1\Delta_u$ electronic state of the linear molecule splits upon bending into a B_u and a A_u component (the Renner-Teller effect). The B_u component retains also at *trans*-bent geometries the composition given by the first of Eqs. (24), on the other hand the 2^1A_u species become dominant in the A_u component of the $^1\Delta_u$ state. The composition of the two lowest-lying singlet electronic states of A_u symmetry can be interpreted as mixing of the $^1\Sigma_u^-$ and $^1\Delta_u$ states at *trans*-bent geometries.

The mixing between the B_u component of the $^1\Delta_u$ state and the species of the same symmetry (B_u) correlating with the $1\Sigma_u^+$ state of the linear molecule is not significant (at least at small distortions from linearity), because of a relatively large energy difference between the $^1\Sigma_u^+$ and $^1\Delta_u$ states.

The fact that the state correlating with $^1\Sigma_u^-$ corresponds to the excitation from the $1a_u$ orbital, the energy of which does not change significantly upon bending, into $4a_g$, being stabilized at bent geometries (see Fig. 1a), has as a consequence the geometry of this electronic state being bent at equilibrium. On the other hand, the A_u component of the $^1\Delta_u$ state corresponds to the excitation from the $3b_u$ orbital, the energy of which decreases upon bending, into $1b_g$, with an energy practically independent of variation in the geometry; this leads to a continuous increase of the energy for the electronic state in question with increasing distortion from linearity.

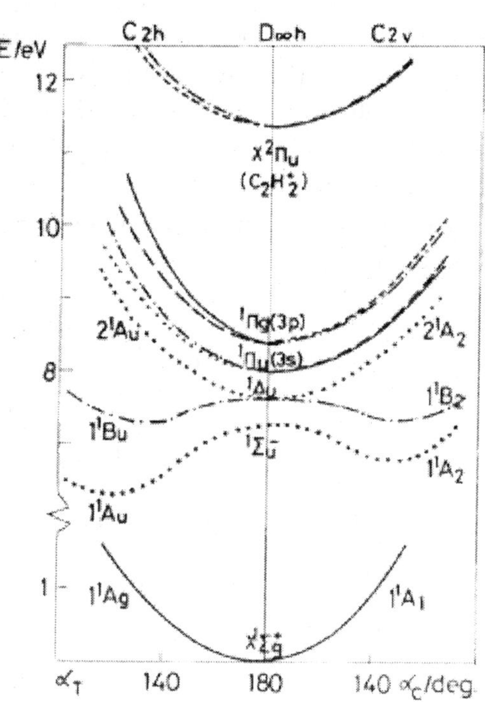

Fig. 2. Trans (left-hand side) and cis- (right-hand side) bending potential curves for singlet electronic states of acetylene.

The other component of the $^1\Delta_u$ state (of B_u symmetry) is described by two leading configurations corresponding to $3b_u \rightarrow 4a_g$, $1a_u \rightarrow 1b_g$ excitations with respect to the ground state. Inspection of Fig. 1a shows that it is difficult to predict precisely whether linear or bent geometry will be preferred by an electronic state of such a composition; the results of explicit *ab initio* computations (Fig. 2) show that it is non-linear, but with a very flat potential curve.

Rydberg electronic states arise by excitations out of the orbitals 1au and $3b_u$, correlating with π_u, into the MOs involving Rydberg-type AOs. Since the latter show negligible dependence on the molecular geometry, all Rydberg states possess linear equilibrium geometry. The states arising by excitations out of the $1a_u$ MO lie below those corresponding to excitations from its lower-energy counterpart $3b_u$, as, for example, the A_u component of the first singlet Rydberg state, $1^1\Pi_u$ [$1a_u \rightarrow 3s_R(a_g)$ excitation with respect to the B_u component [$3b_u \rightarrow 3s_R(a_g)$ excitation of the same state (see Fig. 2).

At *cis*-bent molecular geometries (point group C_{2v}), the π_u and π_g orbitals split into $1b_1+4a_1$ and $3b_2+1a_2$, respectively. One-electron excitations out of $1b_1$ ($\equiv b_1$) or $4a_1$($\equiv a_1$) into $3b_2$ ($\equiv b_2$) or $1a_2$ ($\equiv a_2$) lead to the electronic species represented by the following sixteen Slater determinants

$$D_1^C = |a_1\alpha, b_2\alpha|, D_2^C = |a_1\alpha, b_2\beta|, D_3^C = |a_1\alpha, a_2\alpha|, D_4^C = |a_1\alpha, a_2\beta|,$$
$$D_5^C = |a_1\beta, b_2\alpha|, D_6^C = |a_1\beta, b_2\beta|, D_7^C = |a_1\beta, a_2\alpha|, D_8^C = |a_1\beta, a_2\beta|,$$
$$D_9^C = |b_1\alpha, b_2\alpha|, D_{10}^C = |b_1\alpha, b_2\beta|, D_{11}^C = |b_1\alpha, a_2\alpha|, D_{12}^C = |b_1\alpha, a_2\beta|, \quad (25)$$

$$D_{13}^C = |b_1\beta, b_2\alpha|, D_{14}^C = |b_1\beta, b_2\beta|, D_{15}^C = |b_1\beta, a_2\alpha|, D_{16}^C = |b_1\beta, a_2\beta|.$$

The spectroscopic states built by linear combinations of the determinants (25) are

$$
\begin{aligned}
1\,^1A_2 &= \frac{1}{\sqrt{2}}(D_4^C - D_7^C) = \frac{1}{\sqrt{2}}(a_1a_2 + a_2a_1)\,\Theta^1,\\
2\,^1A_2 &= \frac{1}{\sqrt{2}}(D_{10}^C - D_{13}^C) = \frac{1}{\sqrt{2}}(b_1b_2 + b_2b_1)\,\Theta^1;\\
1\,^3A_2\,(M_s = 1) &= D_3^C = \frac{1}{\sqrt{2}}(a_1a_2 - a_2a_1)\,\Theta_1{}^3,\\
1\,^3A_2\,(M_s = 0) &= \frac{1}{\sqrt{2}}(D_4^C - D_7^C) = \frac{1}{\sqrt{2}}(a_1a_2 - a_2a_1)\,\Theta_0{}^3,\\
1\,^3A_2\,(M_s = -1) &= D_8^C = \frac{1}{\sqrt{2}}(a_1a_2 - a_2a_1)\,\Theta_{-1}{}^3;\\
2\,^3A_2\,(M_s = 1) &= D_9^C = \frac{1}{\sqrt{2}}(b_1b_2 - b_2b_1)\,\Theta_1{}^3,\\
2\,^3A_2\,(M_s = 0) &= \frac{1}{\sqrt{2}}(D_{10}^C + D_{13}^C) = \frac{1}{\sqrt{2}}(b_1b_2 - b_2b_1)\,\Theta_0{}^3,\\
2\,^3A_2\,(M_s = -1) &= D_{14}^C = \frac{1}{\sqrt{2}}(b_1b_2 - b_2b_1)\,\Theta_{-1}{}^3,\\
1\,^1B_2 &= \frac{1}{\sqrt{2}}(D_2^C - D_5^C) = \frac{1}{\sqrt{2}}(a_1b_2 + b_2a_1)\,\Theta^1,\\
2\,^1B_2 &= \frac{1}{\sqrt{2}}(D_{12}^C - D_{15}^C) = \frac{1}{\sqrt{2}}(b_1a_2 + a_2b_1)\,\Theta^1,\\
1\,^3B_2\,(M_s = 1) &= D_1^C = \frac{1}{\sqrt{2}}(a_1b_2 - b_2a_1)\,\Theta_1{}^3;\\
1\,^3B_2\,(M_s = 0) &= \frac{1}{\sqrt{2}}(D_2^C + D_5^C) = \frac{1}{\sqrt{2}}(a_1b_2 - b_2a_1)\,\Theta_0{}^3;\\
1\,^3B_2\,(M_s = -1) &= D_6^C = (a_1b_2 - b_2a_1)\,\Theta_{-1}{}^3;\\
2\,^3B_2\,(M_s = 1) &= D_{11}^C = \frac{1}{\sqrt{2}}(b_1a_2 - a_2b_1)\,\Theta_1{}^3;\\
2\,^3B_2\,(M_s = 0) &= \frac{1}{\sqrt{2}}(D_{12}^C + D_{15}^C) = \frac{1}{\sqrt{2}}(b_1a_2 - a_2b_1)\,\Theta_0{}^3;\\
2\,^3B_2\,(M_s = -1) &= D_{16}^C = \frac{1}{\sqrt{2}}(b_1a_2 - a_2b_1)\,\Theta_{-1}{}^3.
\end{aligned}
\tag{26}
$$

The MOs in the $D^{\infty h}$ point group are connected with those of the C_{2h} by the relations

$$\pi_u = \frac{1}{\sqrt{2}}(a_1 + i\,b_1),\ \ \bar{\pi}_u = \frac{1}{\sqrt{2}}(a_1 - i\,b_1)$$
$$\pi_g = \frac{1}{\sqrt{2}}(b_2 + i\,a_2),\ \ \bar{\pi}_g = \frac{1}{\sqrt{2}}(b_2 - i\,a_2). \tag{27}$$

The states of the $D_{\infty h}$ and C_{2v} point groups correlate with one another in the following way:

$$\frac{1}{\sqrt{2}}(^1\Delta_u + {}^1\bar{\Delta}_u) \rightarrow \frac{1}{\sqrt{2}}(1^1B_2 - 2^1B_2),$$
$$\frac{1}{\sqrt{2}}(^1\Delta_u - {}^1\bar{\Delta}_u) \rightarrow \frac{i}{\sqrt{2}}(2^1A_2 + 1^1A_2),$$
$$\frac{1}{\sqrt{2}}(^3\Delta_u + {}^3\bar{\Delta}_u) \rightarrow \frac{1}{\sqrt{2}}(1^3B_2 - 2^3B_2),$$
$$\frac{1}{\sqrt{2}}(^3\Delta_u - {}^3\bar{\Delta}_u) \rightarrow \frac{i}{\sqrt{2}}(2^3A_2 + 1^3A_2),$$
$$^1\Sigma_u^+ \rightarrow \frac{1}{\sqrt{2}}(2^1B_2 + 1^1B_2), \tag{28}$$
$$^1\Sigma_u^- \rightarrow \frac{i}{\sqrt{2}}(1^1A_2 - 2^1A_2),$$
$$^3\Sigma_u^+ \rightarrow \frac{1}{\sqrt{2}}(1^3B_2 + 2^3B_2),$$
$$^3\Sigma_u^- \rightarrow \frac{i}{\sqrt{2}}(1^3A_2 - 2^3A_2).$$

Again only singlet electronic states will be discussed. The energy of the state correlating with the X $^1\Sigma_g^+$ linear species, ... $1b_1^2 4a_2^2$, increases upon *cis*-bending. In analogy with the situation at *trans*-bending, the first excited singlet state ($^1\Sigma_u^-$ at linear geometry) is at *cis*-bent geometries predominantly dèscribed by the single 1^1A_2 wave function ($1b_1 \rightarrow 3b_2$ excitation with respect to the ground state), and the component of the $^1\Delta_u$ state of the same symmetry by 2^1A_2 ($4b_1 \rightarrow 1a_2$ excitation). The other ^1A_u component (of B_2 symmetry) retains the composition it has at linear geometry ($4a_1 \rightarrow 3b_2$, $1b_1 \rightarrow 1a_2$). Reasoning analogous to that carried out for trans-bending leads to the conclusion that the first singlet state of A_2 symmetry (correlating with $1\Sigma_u^-$) has bent equilibrium geometry, while the second A_2 species (which correlates with $^1\Sigma_u^-$) prefers linear geometry.

The lowest-lying B_1 state (the other $^1\Delta_u$ component) is slightly non-linear, like its B_u *trans*-planar counterpart. All Rydberg-type species are predicted to be more stable at linear geometry than at cis-planar nuclear arrangements.

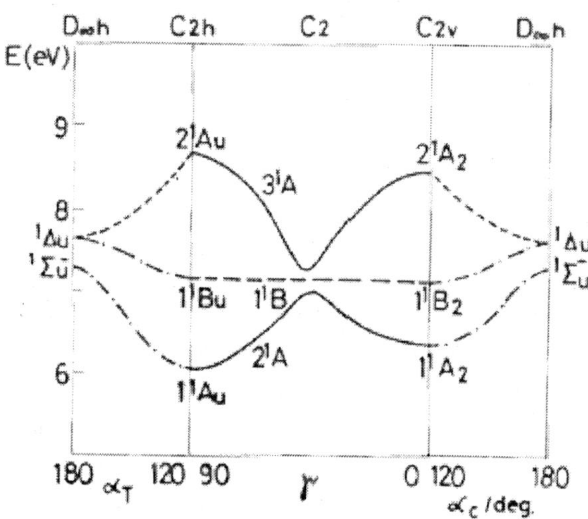

Fig. 3. Torsional potential curves for low-lying excited singlet states of acetylene corresponding to a H–C–C bond angle of 120°.

Torsional Potential Curves

The dependence of the energy of the MOs on the torsional angle, presented in Fig. 1b, is directly reflected in the form of the potential curves for torsional motion. The lowest- energy potential curve corresponds to the electronic configuration ... $3b^2\ 4a^2$, 1^1A. This state corresponds at trans-planar geometry to 1^1A_g and at *cis*-planar nuclear arrangement to 1^1A_1, both of these species correlating with the $X^1\Sigma_g^+$ state of the linear molecule. A consequence of the very weak torsional dependence of all the MOs involved in the wave function of the 1^1A state is that the corresponding potential curve has the form of a nearly straight line.

On the other hand, the composition and energy of the next two 1A electronic states in order of increasing energy show dramatic changes upon torsion. The lower-energy one of them correlates at trans-planar geometry with the ... $3b_u^2\ 1a_u\ 4a_g$, 1^1A_u species and at cis-planar geometry with ... $1b_1\ 4a_1^2\ 3b_2$, 1^1A_2. This means that at large values of the torsional angle γ (i.e., at relatively small torsional distortions with respect to the trans-planar geometry) the electronic configuration for the 2^1A electronic state is ... $3b^2\ 4a\ 5a$ ($4a \rightarrow 5a$ electronic excitation with respect to the 1^1A state), and at small γ values (nearly cis-planar geometry) it is ... $3b\ 4a^2\ 4b$ ($3b \rightarrow 4b$ electronic excitation with respect to the 1^1A state).

This change in the composition of the wave functions for the 2^1A adiabatic state is easily understandable in terms of the MO diagram presented in Fig. 1b: At large γ values, the $4a \rightarrow 5a$ excitation is energetically more favorable than $3b \rightarrow 4b$, while the situation is opposite for small values of γ. The behavior of the 3^1A state is complemental to that of its 21A counter part. This leads to an "avoided crossing" of the 2^1A and 3^1A adiabatic potential side: *cis*-bending curves. curves and consequently to a potential barrier for the 2^1A state at $\gamma \approx \pi/4$.

On the other hand, the torsional potential curve for the 1^1B state, connecting with each other the 1^1Bu (C_{2h} geometry) and 1^1B_2 (C_{2v}) states, has a relatively monotonous form. This concerns also all the Rydberg-type electronic states. In Fig. 3 are displayed the ab initio computed torsional curves for the 2^1A, 31A and 1^1B electronic states. They confirm the above analysis. It should be noted, however, that ab initio calculations showed that the electronic states in question are at non-planar geometries appreciably admixed by Rydberg-type MOs.

Electronic Spectra of Acetylene

At linear nuclear arrangement, representing the equilibrium geometry of the ground state $X^1\Sigma_g^+$ of acetylene, electronic transitions to all low-lying valence-type excited states and $^3\Sigma_u^+$, $^3\Delta_u$, $^3\Sigma_u^-$, $^1\Sigma_u^-$ and $^1\Delta_u$ are forbidden. The lowest-energy dipole allowed transition in absorption involves the first member of the singlet Rydberg series, $^1\Pi_u(3_{sR})$. For this reason the majority of the experimental studies have been devoted to the investigation of the Rydberg spectrum of acetylene. However, when the linear molecular geometry is distorted, many of the "vertically forbidden" transitions become allowed.

The first excited singlet state ($^1\Sigma_u^-$ at linear geometry) correlates with the 1^1A_u species at C_{2h} geometry and with 1^1A_u at C_{2v}. While the electronic transition from the ground state (1^1Ag at C_{2h}, 1^1A_1 at C_{2v}) to the latter species remains forbidden, the $1^1A_g' \rightarrow 1^1A_u$ transition at trans-planar geometry is allowed. However, such a "non-vertical" transition is of low intensity. The spectrum arising from a transition into the 1^1B_u state, correlating at the linear geometry with $^1\Delta_u$ has also been observed. It is possible that some of the features ascribed to this spectrum originate from the transition into the B_2 component of the $^1\Delta_u$ state. On the other hand, the transition from the ground state into the A_2 component of $^1\Delta_u$ is forbidden, and the spectrum involving the A_u component of $^1\Delta_u$, preferring linear geometry, should be extremely weak.

10

Structure and Bonding in Main Group Chemistry

When an element forms a chemical compound, electrons are either lost, gained or shared with other atoms. These tendencies can be assessed by the parameters of ionization energy (IE), electron affinity (EA) and electronegativity. Prediction of bond types either as ionic or covalent allows prediction of the chemical and physical properties of chemical substances. IE refers to the loss of an electron from a gaseous atom or ion (equation 1). Successive loss of electrons from an atom becomes increasingly difficult (because the resulting positive ion holds on to its remaining electrons even more strongly), so, for example, third IEs are always higher than second IEs, which in turn are higher than first.

$$M^{n+}_{(g)} \rightarrow M^{(n+1)+}_{(g)} + e- \quad (1)$$

Going down a group, IEs decrease. Atoms increase in size and the electron to be removed is further from the nucleus; although the nuclear size is much increased, outer electrons are shielded by completed, filled, inner shells, so the effective nuclear charge felt by an outer electron is much less. Crossing the Periodic

Table, first IEs increase because extra protons are being added to the nucleus, and electrons are being added to the same electron shell (Figure 1). These electrons are not very efficient at screening each other from the nuclear charge, so they are attracted more strongly by the nucleus and are harder to ionize. Removal of an electron from a filled shell requires a large amount of energy: the first IE of neon is very high in comparison to the next element, sodium, where loss of an electron will leave a filled shell; Group 1 metals therefore have low first IEs but very high second IEs.

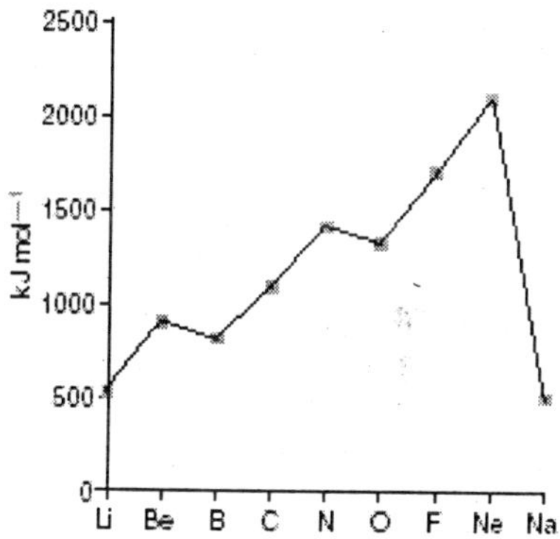

Fig. 1. First ionisation energies for the elements Li to Na

Superimposed on the general trend of increased IE with atomic number within a period, are kinks at boron and oxygen. For beryllium, the 2*s* level is filled, so going to boron involves adding an electron to one of the 2*p* orbitals. Despite the increase in nuclear charge there is a decrease in IE because of the relatively efficient shielding of the 2*p* electron by the 2*s* electrons. At nitrogen, the three 2*p* orbitals each contain one electron (Hund's rule), so going to oxygen involves pairing an electron in one of the 2*p* orbitals. The two electrons in the same orbital repel each other, so the first IE of oxygen is lower than that of nitrogen, because loss of one of the paired electrons is assisted by electron-electron repulsions.

Problem

Figure 2 shows the variation in second ionization energy for the elements Li to Na. Comment on the shape of the graph by comparison with first ionization energies of the same elements (Figure 1).

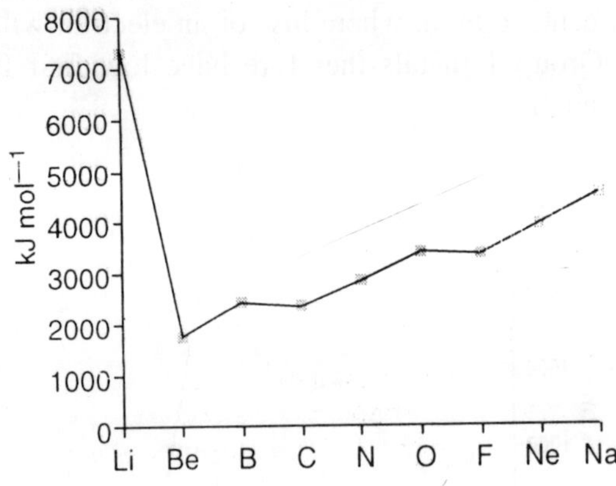

Fig. 2. Second ionisation energies for the elements Li to Na

The same general trend is observed, with a general increase in second IE with increasing atomic number; for a given element the second IE is much higher than the first IE because it involves removal of an electron from a positive ion. The lowest second IE is for Be, since this involves the loss of the second 2s electron, leaving a filled $1s^2$ shell, and the highest second IE is for Li, since this involves loss of one of the 1s electrons (the sole 2s electron is lost in the first IE of Li). The kinks in the graph of second IE are similar of those in the first IE graph, except that the 'pattern' is shifted one element to the right. Thus, the decrease in first IE which occurs from N to O occurs in the second IE graph from O to F, because the process involves the loss of the same electrons.

Electron affinity (EA) is defined as the energy change on addition of an electron to a gaseous atom or ion (equation 2).

$$X^{n-}_{(g)} + e^{-} \rightarrow X^{(n+1)-}_{(g)} \qquad (2)$$

EAs are most negative for elements in the top right of the Periodic Table (i.e. the halogens), while second and higher EAs are always positive because it is more difficult to add an electron to an already negatively charged ion. EAs also become positive on adding electrons to a new shell, so the first EAs of oxygen, fluorine and neon are -141, -322 and +29 kJ mol^{-1}, respectively. Trends in EA are generally less simple than trends in IE. When we consider IEs and EAs together, elements in the bottom-left corner of the Periodic Table have low IEs and EAs and readily lose electrons to form cations, whereas elements in the top-right corner (the halogens, oxygen and sulfur) have high IEs and large negative EAs and readily gain electrons to form anions.

Elements in the middle (particularly the lighter elements) have intermediate IEs and EAs and generally form covalent bonds in their compounds. However, unequal electron sharing results in polar bonds, and this is best discussed in terms of atom electronegativities. Electronegativity refers to the tendency of an atom, in a molecule, to attract electrons to itself.

A scale of electronegativity was devised by Linus Pauling, based on bond energies. While several other electronegativity scales have been developed, the one by Pauling is still widely used. The most electronegative elements are in the top right of the Periodic Table, with fluorine being the most electronegative with the maximum value of 4.0 on the Pauling scale. Electronegativity is a useful general parameter for predicting the general chemical behaviour of an element, and gives good indications of bond types. In general terms, two elements with a large electronegativity difference will tend to form ions, though smaller electronegativity differences are needed when one of the compounds is a highly electropositive metal (e.g. of Group 1). Two elements with similar and intermediate electronegativities (around 2.5) will tend to form covalent] compounds. This is illustrated by C and H, which form an extensive range of covalently bonded organic compounds.

Periodic Trends among the Main Group Elements

The main group elements, and their chemical compounds, cover a wide range of bonding types, from ionic, through polymeric, to molecular. In this section we will survey the general features of the chemistry of the main group elements and selected compounds, using the variation in electronegativity of the elements as a qualitative tool for rationalizing the features. The compounds surveyed are the hydrides, oxides and chlorides, which are some of the most important compounds, and which illustrate the general features very well.

The Elements

It is noteworthy that the p block is the only part of the Periodic Table to contain non-metallic elements. There is a general trend from metallic elements at the bottom left of the Periodic Table (the s-block metals) to non-metallic elements at the top right of the Table (the halogens and noble gases) (Figure 3). This correlates very well with the electronegativities of the elements. Metals are good conductors of heat and electricity, and in solid metals the electrons are extensively delocalized over the whole material. Non-metallic elements are insulators and have no delocalized bonding, instead being formed from localized covalent bonds. In the centre of the p block, there are also so-called metalloid elements such as boron and silicon, which show intermediate electronegativities; they also show relatively low electrical conductivity (compared to metals), but it increases with temperature. The change in properties is nicely illustrated by looking at the first long period, Na to Ar. Na and Mg are both electropositive metals; the next element, aluminium, is a metal, but shows several characteristics of non-metals in forming many covalent compounds. In Group 14, silicon is a metalloid, the element being a semiconductor, and has compounds which show characteristics of both metal compounds and non-metal compounds. By the time we get to phosphorus in Group 15, we are truly in the domain of non-metals; phosphorus exists in several elemental forms, all of which contain covalent P-P bonds. In Groups 16 (sulfur) and 17 (chlorine) the elements are also true non-metals, sulfur existing as covalent S_8 rings (and other forms), and chlorine forming diatomic, covalently bonded molecules.

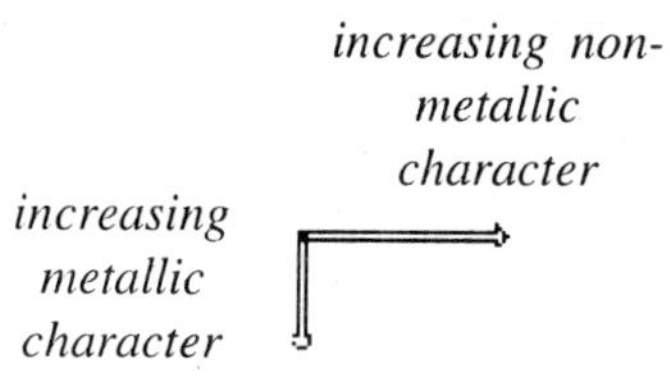

Fig. 3.

Variation in metallic and non-metallic character in the periodic table

Argon exists as a monoatomic gas under ambient conditions, and does not participate in chemical bonding owing to the filled valence shell of electrons, and high ionization energy. Going down any of the main groups, elements become more metallic in character, paralleled by a decrease in electronegativity.

Main Group Element Hydrides

The properties of the main group element hydrides range from ionic (for the s-block metals, with the exception of beryllium), through polymeric (AlH_3), to molecular covalent hydrides for the elements of Groups 14-17. In Groups 1 and 2, the metals are less electronegative than hydrogen (Pauling scale electronegativities: Na 0.9, H 2.1), so the bonding in the hydrides of these metals is predominantly ionic, as M^+H^-; these hydrides react violently with water, generating H_2 gas. For boron and beryllium, the electronegativity diference between the element and hydrogen is small. BeH_2 is covalent and boron hydrides are covalent clusters. In Group 14 the hydrides are all covalent molecular species, typified by CH_4. Continuing across the

Periodic Table, to Groups 15, 16 and 17, the hydrides are all molecular covalent species, with acidity in aqueous solution increasing on moving to the right, as the electronegativity difference between the element and hydrogen increases and the H-X bond becomes more polarized: $H^{\delta+}$–$X^{\delta-}$.

Main Group Element Chlorides

Like the hydrides, the properties of the chlorides follow a broadly similar pattern, with chlorides of metals being ionic and of non-metals being covalent molecular in structure. Thus, for the Group 1 and 2 metals (except beryllium) the chlorides are ionic solids which form neutral solu-Worked Problem 1.2 Q Predict the properties of the hydrides formed by elements with electronegativities of (a) 0.9 and (b) 3.5. A (a) An element with an electronegativity of 0.9 is a metal; it will probably form an ionic hydride which will react with water, giving hydrogen and a basic solution of the hydroxide. (b) This element is a non-metal, and the hydride will be covalent with a polar H-X bond and will dissolve in water, probably giving a neutral to acidic solution. tions in water. The chlorides of small, highly polarizing metal ions such as beryllium, aluminium, gallium and some other elements are polymeric in the solid state.

The majority of the chlorides of the Groups 14 and 15 elements, and BCl_3, are molecular covalent species. The chlorides of the p-block elements and beryllium generally give acid solutions in water, because they react with it rather than simply dissolving.

Main Group Element Oxides

For main group oxides, there is a similar trend from ionic oxides for the bottom left elements, through polymeric oxides in the centre (many of which are amphoteric), to molecular covalent oxides for the elements of higher electronegativity on the right-hand side of the p block. Oxygen is the second most electronegative element, so in combination with (low electronegativity) Group 1 and 2 metals at the left-hand side of the Periodic Table, the resulting oxides are ionic. Examples include Na_2O and CaO. Such oxides are basic oxides, giving highly alkaline solutions in water (equation 3). On moving to the right, to Group 13, the oxides such as B_2O_3 and Al_2O_3 are polymeric and Al_2O_3 is amphoteric. In Group 14 the oxides of the lightest element, carbon, such as CO and CO_2, are molecular oxides; in marked contrast, SiO_2 is a polymeric oxide.

CO_2 is an example of an acidic oxide, since it dissolves in water giving an acidic solution. In Groups 15 and 16 the oxides of nitrogen are all molecular covalent species, many of which are acidic, while those of sulfur (SO_2 and SO_3) are both acidic oxides (equation 4). Likewise, in Group 17, and for xenon in Group 18, the oxides are molecular species.

$$Na_2O_{(s)} + H_2O_{(l)} \rightarrow 2Na^+_{(aq)} + 2OH^-_{(aq)} \quad (3)$$

$$SO_{3(s)} + H_2O_{(l)} \rightarrow 2H^+_{(aq)} + SO_4^{2-}{}_{(aq)} \quad (4)$$

Valence Shell Electron Pair Repulsion Theory

A cursory inspection of the compounds formed by the p-block elements in subsequent chapters reveals that many structures are observed. Even for a certain fixed number of groups around a central atom, there are often different geometrical ways of arranging these; for example, five-coordinate species may be either trigonal bipyramidal or square pyramidal. The simplest and most widely practised method for shape prediction is valence shell electron pair repulsion theory (VSEPR), originally developed in the 1960s, and recently redeveloped by Gillespie. Prediction (or ideally knowledge) of molecular shapes is important for prediction of properties dependent on molecular shape, for example boiling points. The knowledge of bond polarity, determined using the concept of electronegativity, is also important.

Basic Principles of VSEPR

The basic premise of VSEPR is that pairs of electrons in the valence shell of the central atom of a molecule repel each other and take up positions as far apart as possible. The core electrons, which cannot easily be polarized, are conveniently ignored. The shape of a molecule thus condenses to a simple geometrical points-on-a-sphere model; and the basic shapes adopted by molecules with between two and six pairs of electrons on the central atom are given in Table 1. It is important to note that the shape of a molecule or ion can be predicted without knowing anything about the bonding in that species.

Table 1. Shapes of molecules and ions

Number of central atom electron pairs	*Bonding pairs pairs*	*Non-bonding*	*Shape*	*Example*
2	2	0	Linear	$BeCl_2$
3	3	0	Triangular	BF_3
3	2	1	Bent	$SnCl_2$
4	4	0	Tetrahedral	CCl_4
4	3	1	Pyramidal	NH_3
4	2	2	Bent	H_2O
5	5	0	Trigonal bipyramidal (tbp)	PF_5
5	4	1	Pseudo-tbp	BrF_4^+, SF_4
5	3	2	T-shaped	BrF_3
5	2	3	Linear	XeF_2
6	6	0	Octahedral	SF_6, PF_6^-
6	5	1	Square pyramidal	IF_5
6	4	2	Square planar	XeF_4, IF_4^-

Molecules Containing Non-bonding Pairs of Electrons

A bonding pair is shared by two atoms whereas a lone pair is only held by one atom. A lone pair therefore occupies more space in the valence shell of the atom to which it belongs, and it will exert a larger repulsive influence on the other pairs of electrons on that atom. Therefore, in general terms:

— lone pair-lone pair repulsion > lone pair-bonding pair repulsion > bonding pair-bonding pair repulsion

— This is best illustrated by two worked examples: ammonia, NH_3, and water, H_2O.

Molecules with Multiple Bonds

A double bond contains both s and p components. For the purposes of VSEPR, both can be considered to point in the same direction, and we can thus treat a double or triple bond as one superpair of electrons, the effect of which is rather similar to that of a lone pair. As an example, consider the molecule COF_2. The molecule will be trigonal, but the greater space occupied by the double bond will make the fluorine atoms move closer together, decreasing the F-C-F bond angle.

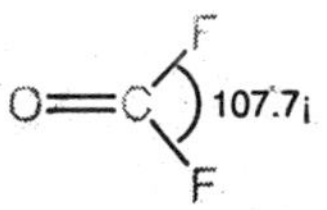

Molecules with Five Electron Pairs

In a metrahedron or octahedron, all of the vertices are identical; however, this is not the case for a trigonal bipyramid, where there are two different types of vertex: axial and equatorial. This is best illustrated by an example, that of PCl_5 in the gas phase, shown in Figure 4. The molecule is a regular trigonal bipyramid. This has implications for molecules which contain five electron pairs, with one or more lone pairs, since there will be a choice of putting the lone pair(s) in axial or equatorial positions. However, on considering the various repulsions in such species, it can be concluded that: lone pairs of electrons, or multiple bonds, always adopt equatorial positions in trigonal bipyramids, owing to the greater space occupied in the valence shell of the central atom. This can be illustrated by species such as SF_4, BrF_4^+, ClF_3 and XeF_2.

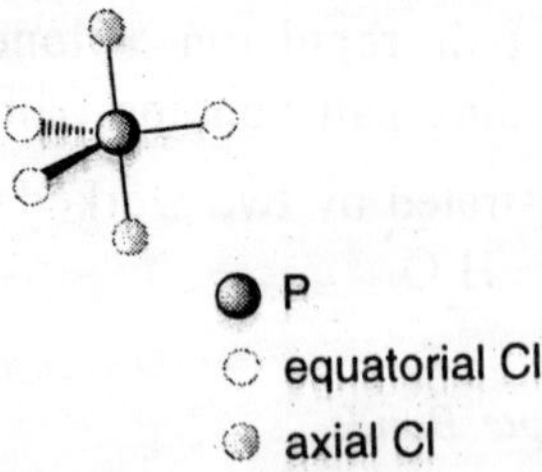

Fig. 4. The structure of gas-phase PCl_5, showing the presence of Cl atoms in axial and equatorial positions

Molecules and Ions with Seven or More Electron Pairs

When seven electron pairs are present in the valence shell of the central atom, the shape is more difficult to predict; there are often a number of different arrangements with similar energies. The three most important regular shapes are the monocapped octahedron, the monocapped trigonal prism and the pentagonal bipyramid, shown in Figure 6.

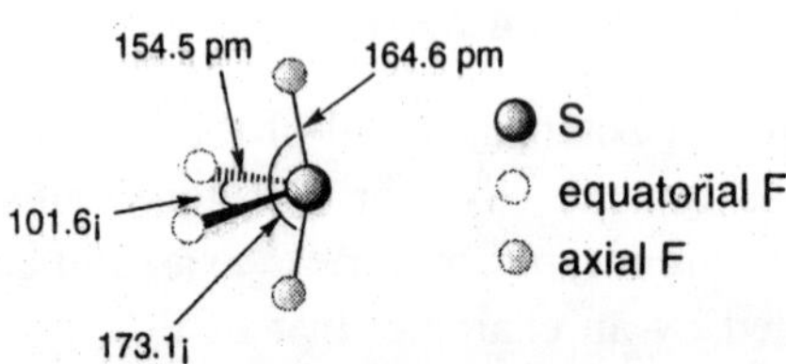

Fig.5. The structure of gas-phase SF_4, showing the presence of F atoms in axial and equatorial positions.

In certain cases, for the heavy p-block elements such as selenium, tellurium, bromine and bismuth, lone pairs occupy spherical s-orbitals, which do not influence the geometry of the species. As an example, $TeCl_6^{2-}$ has one lone pair and six Cl atoms, but it is a regular octahedron; $SeCl_6^{2-}$ and BrF_6 are also octahedral for the same reason. An example of a species with eight electron pairs is XeF_8^{2-}, which has a square antiprism shape, shown in Figure 1.7.

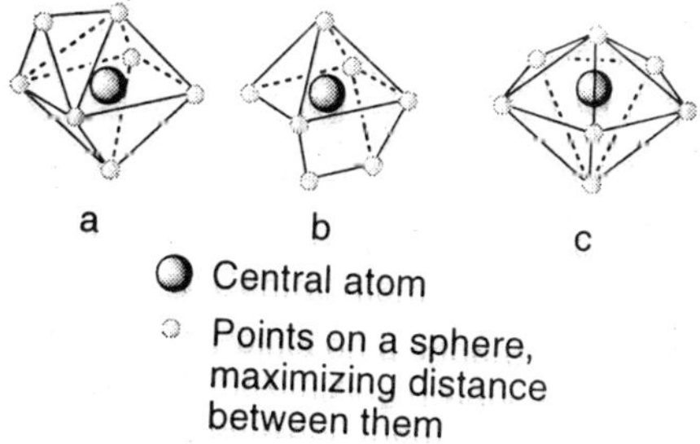

Fig. 6. Common shapes adopted by seven-electron pair species: (a) monocapped octahedron; (b) monocapped trigonal prism; (c) pentagonal bipyramid.

Resonance

When there are two or more resonance forms for a molecule or ion, it is essential that these are considered before predictions of bond angles are made. This can be illustrated by the example of the carbonate ion, CO_3^{2-}. This ion has six electrons (three pairs) around the carbon atom, and so is trigonal.

Fig. 7. The square antiprismatic arrangement of eight fluorines around a central xenon atom in the ion XeF_8^{2-}.

Dative Bonds

A dative bond is fundamentally identical to a normal two-electron covalent bond except that in our electron book-keeping we consider both electrons in the bond to originate from the same atom.

Atom Electronegativities

In an A-X bond between atoms A and X, as the atom X becomes more electronegative, the bonding pair occupies less space in the valence shell of atom A. In practice, this means that in a related series of compounds, bond angles of the type F-A-F are typically smaller than Cl-A-Cl or Br-A-Br angles. The greater size of a Cl atom compared to an F atom also contributes to the widening of the Cl-A-Cl angle. CH_4 and CF_4 have regular tetrahedral shapes, but CF_2H_2 is a distorted tetrahedron with the F-C-F bond angle smaller than the H-C-H angle.

Molecular Orbital Theory

In molecular orbital (MO) theory having localized orbitals which form bonds between pairs of atoms, we construct molecular orbitals which extend over all atoms in a molecule. Space prevents more than a brief summary of the MO treatment of bonding, and in this section it is intended to illustrate the application of MO theory to main group molecules. The discussion will be illustrated by considering the properties of various O_2 species. We will start by looking at a very simple molecule: dihydrogen, H_2. Each *H* atom has a 1s atomic orbital (AO) available for bonding, and these can interact in two ways. In-phase interaction gives a bonding MO, σ, while out-of-phase interaction gives an antibonding MO, σ*. The bonding MO is symmetrical about the centre of the molecule, and there is an increase in electron density in the internuclear region compared to the two H 1s orbitals; the bonding MO σ therefore has a lower energy than the energy of the hydrogen 1s atomic orbital.

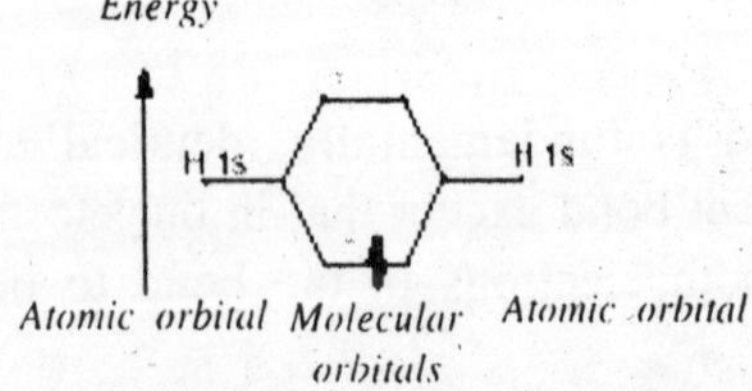

Fig. 8. MO energy level diagram for H2

In contrast, the antibonding MO σ^* has a decrease in electron density in the internuclear region (when occupied by electrons), and is higher in energy than the 1s atomic orbitals. The two electrons (one from each hydrogen atom) enter the bonding MO σ, corresponding to a net H-H single bond. If two more electrons are added to H_2, the antibonding MO σ^* becomes filled. In this case, equal numbers of electrons are present in bonding and antibonding MOσ, there is no net bonding, and H_2^{2-} is unfavoured relative to its dissociation product, two H^- (hydride) anions.

In dioxygen species O_2, the MO scheme is slightly more complex because each O atom has both 2s and 2p AOs available for bonding. The p_z orbitals point towards each other (the molecular axis is defined as the *z* axis) and can interact to form σ bonding (σ_3) and antibonding (σ_4^*) MOs. The p_x orbitals on each oxygen can interact in a side-on manner, to form π bonding (π_1) and antibonding (π_2^*) MOs, respectively. In the same way, the p_y orbitals interact to form π bonding (π_1) and antibonding (π_2^*) MOs which are degenerate with the p_x- derived π MOs. The MO diagram in Figure 1.11 can be used to describe the bonding in O_2^+, O_2, O_2^- and O_2^{2-}. As with atoms, electrons enter MOs from the lowest énergy levels first, and if there are degenerate orbitals, electrons initially singly occupy each with parallel spins.

Table 2 Bond lengths of some dioxygen species

Species	*Name*	*Bond order*	*Bond length (pm)*
O_2^+	Oxygenyl	2.5	112.3
O_2	Dioxygen	2.0	120.7
O_2^-	Superoxide	1.5	128
O_2^{2-}	Peroxide	1.0	149

For O_2 it can be seen that the π^{2*} MO contains two unpaired electrons, and hence O_2 is predicted to be paramagnetic and attracted by a magnetic field. This fits very well with the physical properties of O_2, which is indeed paramagnetic. It is important to note that a simple valence bond description of O_2 (as O=O) does not predict any unpaired electrons, and this therefore represents one of the major triumphs of MO theory.

The total bond order for O_2 from the MO diagram is 2, in accordance with the simple valence bond description of O=O. The oxygenyl cation, O_2^+, has one less electron in the antibonding π^{2*} MO, so the bond order is 2.5. Similarly, O_2^- and O_2^{2-} have respectively one and two more electrons in π^{2*}, so the bond orders are 5 and 1, respectively. The predicted bond orders correlate very well with the experimental bond lengths, with a larger bond order giving a shortened bond, as shown in Table 2. Similarly, O_2^+, O_2 and O_2^- contain unpaired electrons and are paramagnetic, while O^2_{2-} has no unpaired electrons and is diamagnetic.